CHATGPT
FOR
CONTENT MARKETING
SUCCESS

CHATGPT FOR CONTENT MARKETING SUCCESS

A STEP-BY-STEP GUIDE FOR PROFESSIONAL SERVICE COMPANY OWNERS TO CREATING CONTENT CONSISTENTLY

DANNI WHITE

DW CREATIVE PUBLISHERS

ChatGPT for Content Marketing Success: A Step-by-Step Guide for Professional Service Company Owners to Creating Content Consistently.

FIRST EDITION

ISBN 978-1-952605-34-5 (print paperback)

ISBN 978-1-952605-35-2 (eBook)

Library of Congress Control Number: 2023941688

Cover design: Julia Kuris of DW Creative Publishers

Book design: Sajjad Haider of DW Creative Publishers

Editing: DW Creative Publishers

Adapted for eBook by Sajjad Haider

TABLE OF CONTENTS

CHANGE IS INEVITABLE.

GROWTH IS OPTIONAL.

—JOHN C. MAXWELL

HOW TO USE THIS BOOK

C hatGPT for Content Marketing Success: A Step-by-Step Guide for Professional Service Company Owners to Creating Content Consistently has been thoughtfully created to guide you through the different stages of your content creation, marketing, optimization, and distribution journey using artificial intelligence based tools as your supporting cast member. This book is designed for business owners and entrepreneurs just as much as it is for marketing professionals who want to hone their craft.

We will start by understanding the foundation of ChatGPT, the different types of artificial intelligence and how we got from GPT-1 to GPT-4. Next, we'll delve into the various aspects of effectively leveraging ChatGPT for content marketing. And then we will look at some actual prompts that you can use in ChatGPT to effectively save time and uplevel your content marketing efforts.

This book is a manual, but it is also a strategic resource that offers actionable insights, foundational understanding, and critical strategies you can incorporate immediately into the content creation process of your business. Whether you are a solopreneur, experienced business owner, serial entrepreneur, or someone looking for a way to streamline content marketing efforts, this book is for you. The real value in these pages is the potential to save you time and effort in content creation while significantly enhancing the quality and consistency of your output.

Use this book as a reference, a guide, and a companion in your content marketing journey. Whether you're drafting a blog post, composing an email campaign, or curating content for social media, we all want to be able to do things much faster in our jobs and this book has valuable advice that can be applied in each scenario.

By adopting the strategies and practices outlined in this guide, you'll be solidifying your knowledge of artificial intelligence, aligning your business to the future of marketing, fostering a culture of innovation, and unlocking new avenues for growth and success in your business.

INTRODUCTION

Welcome to *ChatGPT for Content Marketing Success: A Step-by-Step Guide for Professional Service Company Owners to Creating Content Consistently*. This isn't just a book; it's your front-row ticket to the transformational world of AI-driven content marketing.

In the ever-evolving digital age that we live in, content is more than just a means of conveying information—it is the lifeblood of engagement, the cornerstone of building trust, and a significant driver of business success. However, the world of content marketing is always changing (just as much as the algorithms change within Google search and social media platforms) and it is a fiercely competitive space. Standing out, staying relevant, and consistently creating compelling content that actually matter to your audience and turns into revenue is far easier said than done, especially for professional service company owners who often wear many hats.

Imagine having a tool that could help you navigate this intricate digital landscape with precision and efficiency. Imagine having a companion by your side that could help you enhance your brand presence, create an array of content formats for your audience, and include your that your brand style, tone and voice stand out among the noise.

That's exactly where ChatGPT comes into play.

ChatGPT has been out for a while now, but I think it can be very challenging to understand how something is to be used fully and extensively to cut time on manual processes and labor without spending some time with the tool or technology first and using it in all the ways imaginable. OpenAI's ChatGPT isn't just another AI technology; it's truly a game-changer.

Powered by natural language processing and machine learning algorithms trained on a diverse range of data, ChatGPT brings a unique blend of creativity and scalability to the content marketing (and marketing as a whole) table. Aside from marketing, it is hugely effective for business analysis, market research, and everyday life such as meal planning for a family of four for the month.

How can we harness all of this good power effectively for our businesses and our personal lives? That's exactly what this book aims to answer.

There is a lot of information out there about what ChatGPT can do and how it should be used and whether artificial intelligence is taking over our entire lives. I think it is safe to say that just like other technologies and tools have entered onto the playing field of our everyday lives, it is our responsibility to do more learning and understanding than worrying.

In the next several chapters, we will dive into the nuts and bolts of leveraging ChatGPT to turbocharge our content marketing strategy, creation, distribution, optimization, and measurements. Whether you're writing blog posts, social media content, email marketing campaigns, advertising copy or web copy, this book provides a practical, step-by-step roadmap to craft high-quality content consistently and efficiently using ChatGPT.

This book isn't just about the **how.** It's about understanding the **why**. We will explore why AI has become an indispensable tool in content marketing, and why ChatGPT along with other technologies stand out amidst the growing crowd of AI writing tools.

Furthermore, this book is designed to cater to a wide range of readers. Whether you're a seasoned marketing professional looking to stay updated with the latest trends or a business owner looking to understand how AI can enhance your marketing efforts, there's something in here for you. Each chapter builds upon the previous one, taking you on a journey from the basic foundations to the more advanced applications of ChatGPT in content marketing.

By the end of this book, you'll have a solid grasp of how to incorporate ChatGPT into your content creation process, how to think about artificial intelligence, how to be in control and get what you need out of the AI, and how to use it to stay one step ahead in marketing your business.

This isn't just a journey of learning; it's a journey of transformation. I believe tools like ChatGPT aren't created just to make tasks easier or to save us time or to take away people's jobs—it's more about forcing us to amplify human creativity, empower our own minds and voices to think, and revolutionize the way we communicate.

Are you ready to embark on this pretty cool journey? If your answer is **YES**, then flip the page and let's get started.

BY FAR, THE GREATEST DANGER
OF ARTIFICIAL INTELLIGENCE
IS THAT PEOPLE CONCLUDE TOO
EARLY THAT THEY UNDERSTAND IT.

—ELIEZER YUDKOWSKY

WHAT IS CHATGPT?

Chat GPT, also known in long-form as Generative Pre-training Transformer (GPT), is an AI-powered language model developed by OpenAI. Its development marks a significant leap forward in the field of natural language processing, revolutionizing the way we interact with machine learning technologies.

At its core function, ChatGPT is a tool capable of understanding and generating human-like text based on the contextual input it receives. When we say **understanding** here, we mean that the tool can interpret the context and meaning of words within a given input. For clarity, **generating** means it can produce a relevant, cohesive, and at times, insightful response based on the input or prompt it has been given. It does this by employing the transformer architecture and leveraging its training on a vast amount of diverse internet text, allowing it to handle a wide array of topics and conversations from millions of users at one time.

To better understand how ChatGPT works, it's helpful to break down the components of its name.

The 'GPT' part stands for Generative Pretrained Transformer, as mentioned above. This refers to the type of model it is, known as a transformer model, which is pretrained on a large amount of text data and is generative in that it can generate responses or predictions based on input.

'Chat' signifies the model's conversational capabilities, indicating its main use case is to generate human-like dialogues. However, the application of ChatGPT extends beyond simple chatbot functions. Its ability to generate coherent, contextually relevant copy has opened up a myriad of uses in areas such as blog writing, content creation for various formats, tutoring, coding, and translation among a host of other things.

ChatGPT utilizes a machine learning technique known as deep learning, specifically a variant called transformer neural networks. These networks are designed to understand the context of words and sentences by assessing their relationship with other words in a text. Through this, ChatGPT can produce text that is contextually sound and semantically meaningful.

Underneath the hood of this AI powered vehicle, ChatGPT operates based on a sequence of tokens, which can be as small as one character or as large as one word. It scans these tokens left-to-right, predicting the next token based on the ones it has already seen. In simple terms, if given the phrase "The dog chewed the...", it will attempt to predict the next word, such as "bone." Pretty impressive, right?

It's important to remember that while ChatGPT can generate impressively human-like text, it doesn't understand text in the same way humans do. It doesn't possess human capabilities including consciousness, awareness, or emotion. It also does not know about current events or what happened yesterday or a week ago in the world unless you explicitly tell it. The responses it gives are the result of patterns it has learned during training, not from any inherent, human understanding or personal experience. It can't form opinions, beliefs, or offer any factual information about the world outside of its training data.

OpenAI has made several iterations of GPT. GPT-4 is the most advanced publicly available version at the writing of this book. Each version from GPT-1 to the current GPT-4 has seen an increase in the number of parameters and data fed to the tool. GPT-4 is reported to have approximately 1.76 trillion parameters based on eight models with 220 billion parameters each. These parameters help the model learn and generalize the patterns in the data it has been trained on, which improves performance in generating text.

As a side note, OpenAI will keep iterating and adding data and parameters in its upcoming release. GPT-5 is already in the works and is expected to be released some time in 2024. The big expectation is that it will achieve Artificial General Intelligence (AGI), which is the next-gen version of AI models and systems intended to be smarter than a human being. Stick around on the internet for more to come on that. But before that, there may be a release of a GPT-4.5 that is expected to bring multimodal capability which means the tool will be able to analyze both text and images.

The sheer scale and capabilities of the current ChatGPT-4 have made it a versatile tool in many applications. It's being used to draft emails, write code, create content for text, video, and audio, help with homework, create marketing plans and campaigns, give ideas and suggestions on a topic, and answer customer queries.

The goal is to use it responsibly as it has the potential to generate inappropriate, inaccurate, or misleading content (much like us humans, ehh). This is why I always tell my clients and non-clients alike who want to use ChatGPT for ALL of their needs (quite literally) to use ChatGPT abundantly and often, but to take what it gives you with a grain of salt. It requires editing, voice, tone, style, and emotion – literally – all the things that make you, you and me, me.

In short, ChatGPT is a powerful AI language model. By understanding and generating human-like text based on data and inputs, it's transforming our interaction with AI and opening up amazing possibilities for progress in our businesses that require less time and effort. However, like any tool or technology that gets put in front of us, it's important to remember that it is a tool, and its long-term value heavily depends on how we use it responsibly and effectively.

WHAT YOU NEED TO KNOW ABOUT GPT-1, GPT-2, GPT-3 AND GPT-4

Prior to the current GPT-4 language model, OpenAI released three models in the GPT series leading up to that release: GPT-1, GPT-2, and GPT-3. In this chapter, I want to provide a brief but informative overview of all the GPT models to date.

GPT-1: The first model in the GPT series, introduced in 2018, GPT-1 had 117 million parameters. It introduced the transformer architecture that all GPT models use. This architecture allows the model to pay variable amounts of attention to different words in the input when generating each word in the output. GPT-1 was trained on websites, books, and other data texts and demonstrated a good understanding of grammar, facts about the world, and a decent level of reasoning ability.

GPT-2: GPT-2 provided significant advancements over GPT-1. Released in 2019 and containing 1.5 billion parameters, GPT-2 offered over 10 times more information than GPT-1. It was also trained on a much broader dataset. GPT-2's performance showed that scaling up the model size and training data led to significant improvements in the model's ability to generate coherent and contextually relevant sentences. GPT-2, however, opened up the doors to risks of potential misuse which led up the release and opened a conversation that is ongoing to this day around ethical implications of AI.

GPT-3: Launched in 2020, GPT-3 represented another major leap in data context and scale, with 175 billion parameters, making it over 100 times more powerful than GPT-2. GPT-3 can generate impressively human-like text and perform tasks that require a relatively deep understanding of context, various languages, and even some level of reasoning, often without needing task-specific training data. Despite these major advancements, GPT-3 was not without its limitations. In some cases, it generated incorrect or nonsensical responses, lacked common sense in some scenarios, and was verbose in some of its content outputs.

In between the release of GPT-3 and GPT-4, OpenAI released GPT-3.5. GPT-3.5 is the current free version of ChatGPT. It processes only plain text or text-to-text inputs as some call it and it is much faster than GPT-4 in generating responses. In addition, it is not multi-modal in language like GPT-4 is. The big thing to know in this GPT-3.5 process is that a massive amount of data was rated by human raters to express human emotions and preferences. The results of these ratings were used to fine-tune the large language model (LLM) that powers GPT-4. And this is the version that has become very popular among almost everybody when it was released to the world.

Officially launched on March 14, 2023, GPT-4 stands as the newest and latest addition to the GPT model series. Compared to its predecessor, GPT-3, which was already quite impressive, GPT-4 represents a majorly significant advancement. It is said to have approximately 170 trillion parameters which makes it at least 1000 larger than GPT-3

What are parameters? In the context of artificial intelligence, parameters are the intrinsic numerical values and variables of a model that are learned from data during the training process. They shape the model's understanding and determine the input-output relationship of how processes input data and the resulting production of output data. In addition, parameters drive predictions, classifications, and other output behaviors in response to new inputs.

One of the unique attributes of GPT-4 is its multimodal capabilities, meaning it can process images as inputs just as well as textual prompts. For instance, during the live launch of GPT-4, an OpenAI engineer presented the model with a hand-drawn website design image. Impressively, the model generated functional code for the website.

The improved model exhibits superior understanding of complex prompts, performing at the near equivalent levels of humans across various professional and traditional benchmarks. Plus, it boasts a larger context window and context size, meaning it can remember more data during a chat session. In terms of creativity, reliability, accuracy, and intelligence, GPT-4 seems to hit all of those performance benchmarks.

Access to GPT-4 is exclusive to ChatGPT Plus users for $20 per month. It is important to note that unlike GPT-3.5, there is a usage limit in place on GPT-4.

I am sure there will be future GPT model releases as GPT-5 is reportedly currently in the works already. GPT-4 is redefining what's possible with AI tools and is expected to find applicability across numerous industries.

It is important to keep in mind that while larger and more complex models can perform more impressive tasks, they also raise potential risks and ethical concerns, such as misuse for generating disinformation or difficulties in controlling the model's output. Responsible use and oversight of these models is a significant aspect of development and deployment.

In the meantime, as we use ChatGPT and other AI-based tools, we can put in the work to make sure what we end up sharing with the public via social media, in-person, on our websites, or otherwise is as accurate and factual as possible.

HOW TO NAVIGATE THE CHATGPT INTERFACE

Whether you're using the free or paid version of ChatGPT, its interface is very intuitive and user-friendly with an open chat screen and a left side panel that shows your recent chat history. Paid subscribers do receive a few additional features with the biggest one being that GPT-4 is only available under the paid subscription.

ChatGPT Free Version

The unpaid version of ChatGPT is run by the GPT-3.5 model. It can be accessed by going to **chat.openai.com**. If you're a new user, you'll need to create an account. This is a straightforward process where you're required to provide your email and create a password. Once you've confirmed your email, you can log into the platform.

Once logged in, you'll be taken to the main chat interface. The interface resembles a typical chat application. On the right side of the screen, you'll see a text box. This is where you input your prompts to instruct the platform on what you want to generate.

After typing your prompt, press the 'Send' button or hit enter. ChatGPT will process your input and provide a response in the chat window above. The chat window displays the conversation history, allowing you to see both your prompts and ChatGPT's responses.

You can continue the conversation by typing more prompts into the text box. If you don't like the initial response, hit "Regenerate response" and it will spit out rewritten content based on the initial prompt.

ChatGPT Plus Paid Version

ChatGPT Plus offers several benefits that the free version does not. First, users get access even during peak times and priority access to new features and improvements.

Once you subscribe to ChatGPT Plus, which is a $20 a month subscription, you'll use the same chat interface as in the free version. However, you'll notice faster response times from the AI and an improved flow of conversation.

Subscription to ChatGPT Plus can be done through the OpenAI website. It's a monthly subscription, and you can cancel at any time. After subscribing, you'll get immediate access to the benefits. Interacting with ChatGPT, regardless of the version, is a blend of getting work done, learning new information, and exploring all the possibilities of AI.

HOW TO USE CHATGPT EFFECTIVELY TO CREATE CONTENT FOR YOUR BUSINESS (INCLUDES 50 CONTENT FORMATS TO USE)

ChatGPT is a question and answer engine (much like Google is a search engine). It has the potential to generate high-quality, human-like text which opens up a world of timesaving and cost-cutting opportunities for business owners and marketing professionals.

When we leverage ChatGPT, we streamline and shorten the entire content creation process for ourselves, our companies, and the marketers, copywriters, and researchers on our team. You can ask ChatGPT to generate anything from serious how-to content to HTML code to comedic jokes and creative poems.

More specifically, business owners can use ChatGPT to complete or issue the first iteration of a range of tasks including:

- Blog post ideas
- Social media captions
- Hashtags
- Product descriptions
- Website copy
- Outlines for podcasts

- Video transcriptions

- Persuasive ad copy

- Email campaigns

- Data analysis

- Writing code

- Rewriting existing content

- Rephrasing existing content

- Crafting a marketing strategy

- Developing a content calendar

Best Types of Questions to Ask ChatGPT

The key to making ChatGPT work for you is understanding how to ask it the right questions, so it gives you a compelling and comprehensive response. Here are some types of questions that will generate good responses:

- **Fact Seeker:** Questions with known answers such as "What's the tallest mountain on Earth?" or "Who painted the Mona Lisa?" are excellent for ChatGPT. Its large knowledge base can generate answers to factual questions within seconds.

- **Advice Guru**: Need suggestions? Seek guidance on topics like best productivity hacks or top five book recommendations on a particular subject. Just remember that the advice you receive is general advice, it's not personalized to you.

- **Curious Mind**: Complex concepts can be made easy and simple. This is especially useful for teachers, educators, or learning development specialists who need help crafting lesson plans or presentations.

- **Creative Spark**: When you give ChatGPT a creative prompt to tell a story and create a dialogue, it's got your back.

- **Language Lover:** While ChatGPT can only produce content of any magnitude in the English language at the moment, you can ask it simple questions related to language such as word meanings and how to say a word or phrase in another language.

- **Knowledge Seeker:** Explore new themes, subjects or topics with ChatGPT. Whether for school, to pass a test, or for general knowledge in your everyday work, ChatGPT can assist in helping you learn a new topic.

Keep in mind that while ChatGPT is impressive, it is not infallible. Double-check what it spits out, edit and rewrite where necessary, and take a peek at sources for factual information.

As a side note, ChatGPT-4 does not have the most current or latest up-to-date information about any particular subject because its last training cut-off was in September 2021.

50 Ways to Put ChatGPT to Work For You

Here are 50 practical ways that business owners, entrepreneurs, and marketing professionals can use ChatGPT effectively to create content consistently:

1. **Blog Posts:** ChatGPT can help draft compelling blog posts on a variety of topics related to your business, industry trends, product updates, and more.

2. **Social Media Content:** ChatGPT can generate engaging content for various social media platforms, tailored to the specific nuances of each platform.

3. **Website Copy:** Use ChatGPT to create professional, concise, and engaging copy for your business website, ensuring a good first impression for potential customers.

4. **Product Descriptions:** Leverage ChatGPT to generate detailed and appealing product descriptions, enhancing the appeal of your online product catalog.

5. **Email Campaigns:** Create personalized, engaging email content to nurture leads, communicate with customers, and improve email marketing success.

6. **Press Releases:** Draft professional press releases to announce business milestones, new products, or events, using ChatGPT to ensure a clear, concise message.

7. **FAQs:** Use ChatGPT to create comprehensive FAQs, enhancing customer support by answering common questions about your products or services.

8. **Customer Testimonials:** ChatGPT can help write compelling customer testimonials and case studies, highlighting the value and success of your products or services.

9. **Newsletters:** Regular business updates or industry news can be easily drafted using ChatGPT, ensuring engaging and informative newsletters.

10. **Ad Copy:** Improve the effectiveness of your online advertising with optimized ad copy created by ChatGPT, tailored to your target audience and marketing objectives.

11. **Scripts for Promotional Videos:** ChatGPT can generate scripts for promotional videos, product demonstrations, or explainer videos, creating a narrative that captivates your audience.

12. **Podcast Show Notes:** Maximize the reach and utility of your podcast episodes by using ChatGPT to create detailed show notes.

13. **Webinar Descriptions:** ChatGPT can help you create enticing descriptions for your webinars, attracting more sign-ups.

14. **E-Books:** Businesses can use ChatGPT to create comprehensive e-books, serving as valuable lead magnets or educational resources.

15. **Tutorials and How-to Guides**: Explain your product's usage or create educational content related to your business with detailed guides generated by ChatGPT.

16. **Online Course Content**: If you offer online courses, ChatGPT can help create course materials, lessons, and summaries.

17. **White Papers**: Establish your business as a thought leader in your industry by using ChatGPT to draft informative and well-researched white papers.

18. **LinkedIn Articles**: ChatGPT can create professional and engaging articles for LinkedIn to establish your authority and build your network.

19. **Infographic Text**: Use ChatGPT to create concise and informative text for infographics, enhancing visual content with insightful data points.

20. **Event Invitations**: Generate compelling invitations for business events, webinars, or product launches.

21. **Survey Questions**: Use ChatGPT to design insightful survey questions, helping you gather valuable customer feedback and data.

22. **Quizzes**: Engage your audience with quizzes related to your industry or product. ChatGPT can generate both the questions and results descriptions.

23. **SEO Content**: Optimize your digital presence by using ChatGPT to create content rich in keywords, boosting your SEO strategy.

24. **Chatbot Scripts**: Improve your customer service by programming your chatbot with responses generated by ChatGPT.

25. **Proposal and Report Writing**: Use ChatGPT to draft business proposals or create professional reports, saving significant time and effort.

26. **User Manuals**: ChatGPT can create user-friendly instructions for your products, making it easy for customers to understand and use them.

27. **Brainstorming Sessions**: Use ChatGPT as a creative partner to generate ideas for marketing campaigns, blog topics, product features, etc.

28. **Employee Onboarding Documents**: Create comprehensive and informative onboarding guides for new hires.

29. **Internal Communications**: Draft messages or emails for internal communication, saving managerial time.

30. **Content for Affiliate Marketing**: Generate content for your affiliate marketers, helping them effectively promote your product or service.

31. **Slogan/Tagline Creation**: Need a catchy slogan or tagline? ChatGPT can help brainstorm creative options.

32. **Presentations**: ChatGPT can create the text for slides, speaker notes, and even the script for the presenter.

33. **Business Letters**: From thank you letters to business proposals, ChatGPT can help draft various professional letters.

34. **Job Postings**: Write detailed and engaging job descriptions to attract the right talent.

35. **Translation Assistance**: While not a perfect translator, ChatGPT can provide rough translations of text, which can be useful for understanding the gist of non-native languages.

36. **Meeting Agendas/Minutes**: Save time by using ChatGPT to draft meeting agendas or to summarize meeting minutes.

37. **Forum Posts**: Engage with your online community by creating informative and discussion-worthy posts for online forums.

38. **Sales Scripts**: Train your sales team with effective scripts to handle various customer interactions, created by ChatGPT.

39. **Influencer Outreach Messages**: Craft personalized outreach messages for potential influencers in your industry.

40. **Seasonal Messages**: Use ChatGPT to draft holiday greetings or seasonal promotional messages.

41. **Comment Responses**: Automate responses to user comments on your blog or social media platforms.

42. **Brand Storytelling**: ChatGPT can help in weaving a captivating brand story, enhancing your brand identity.

43. **Market Research**: Use ChatGPT to write questions for conducting market research or competitive analysis.

44. **A/B Test Content**: Use ChatGPT to quickly create variations of your content for A/B testing, helping you understand what resonates with your audience.

45. **Speeches**: Draft speeches for webinars, presentations, or events.

46. **Crowdfunding Campaigns**: Create compelling stories and descriptions for your crowdfunding campaigns.

47. **Content Calendar**: ChatGPT can help draft a content calendar, providing a plan for your content marketing activities.

48. **Memes:** Yes, memes! Use ChatGPT to come up with funny and relatable memes that align with your brand identity.

49. **Crisis Communication:** In times of crisis, use ChatGPT to craft empathetic and transparent communication for your customers and stakeholders.

50. **Policies & Guidelines**: Create clear and comprehensive policies or guidelines for your business, including privacy policies, and return policies.

The many ways in which we can utilize ChatGPT are only limited by our imagination and creativity.

While ChatGPT is a very powerful tool, it is important to remember that it is just that – a tool. It should be used responsibly and ethically while integrating the human experience into the content it produces. Using ChatGPT or any other AI tool out there (and there are dozens of them) doesn't replace human creativity, but it can certainly amplify and optimize it, making it a valuable partner throughout the entire content strategy and creation process.

Like any AI, ChatGPT generates content based on the data it's been trained on and the prompts (which we will touch on further in this book) or instructions it has been given. Essentially, you will get out of it what you put into it. Clear and specific prompts are crucial to achieve desired outputs from anything within the list above or other lists given in this book. In most every case, the content it spits out will need a touch of human editing to align better with your brand voice and specific requirements.

200 WAYS TO USE CHATGPT IN YOUR BUSINESS

The list provided here is by no means an attempt to create an exhaustive list of the ways in which you can use ChatGPT for your business. Technology is to be used to our advantage whenever and wherever possible. This is an attempt to share a list of ways in which you might not have thought about using ChatGPT to the benefit of your business.

As a side note, some of the bullet points below may overlap with the previous list of ways you can use ChatGPT to create content, and that is okay. Hopefully, it will get you thinking outside of the box and effectively using this and other tools you have at your disposal.

1. **Company blog comments**: Manage and respond to comments on your company's blog.

2. **Virtual assistant**: Use as a virtual assistant to answer basic business or product-related questions.

3. **Cross-promotion suggestions**: Generate ideas for cross-promotion with other brands.

4. **Up-sell and cross-sell emails**: Write emails to encourage up-selling and cross-selling.

5. **Product FAQ creation**: Create detailed FAQs for each of your products.

6. **Policy writing**: Assist in drafting company policies or procedures.

7. **Q&A sections**: Write questions and answers for webinars, interviews, or presentations.

8. **YouTube descriptions**: Write engaging and SEO-friendly descriptions for YouTube videos.

9. **Value proposition**: Help define your product or service's value proposition.

10. **Meeting agendas**: Assist in preparing meeting agendas to ensure productivity.

11. **Podcast show notes**: Write detailed show notes for your podcast episodes.

12. **Employee recognition**: Write messages recognizing employee achievements.

13. **Performance review templates**: Create templates for employee performance reviews.

14. **Testimonials request emails**: Write emails requesting testimonials from satisfied customers.

15. **User interface text**: Write clear and concise text for your app or website's user interface.

16. **Screenplay for commercials**: Write scripts for television or online commercials.

17. **App store optimization**: Help with writing app descriptions for optimal app store SEO.

18. **Text for infographics**: Write informative and engaging text for infographics.

19. **Micro-copy**: Generate micro-copy for websites, such as button text or error messages.

20. **Text for memes or GIFs**: Create funny and relevant text for memes or GIFs for social media.

21. **Confirmation emails**: Write polite and clear order confirmation or booking confirmation emails.

22. **Influencer contracts**: Assist in drafting contracts for influencer partnerships.

23. **B2B outreach**: Draft professional B2B outreach emails for collaborations or partnerships.

24. **Sweepstakes rules**: Write clear and comprehensive rules for sweepstakes or competitions.

25. **Pricing page copy**: Write clear, persuasive copy for your pricing page.

26. **Employee handbook**: Assist in writing a comprehensive employee handbook.

27. **SWOT analysis**: Provide a structured template for performing a SWOT analysis.

28. **Coupon codes**: Come up with catchy and memorable coupon code names for promotions.

29. **Reseller agreements**: Assist in drafting reseller agreement terms and conditions.

30. **Caption for event photos**: Write engaging captions for event photos shared on social media.

31. **User stories for software development**: Create user stories to guide software development based on customer needs.

32. **Twitter threads**: Write insightful Twitter threads on industry-related topics.

33. **Partner proposal**: Draft partner proposals for business collaborations.

34. **Client onboarding guides**: Create detailed client onboarding guides to ensure a smooth start.

35. **Website metadata**: Write metadata for your website to improve SEO.

36. **Business contracts**: Assist in drafting basic contracts or contract templates.

37. **MOU drafting**: Help with drafting Memorandums of Understanding for business partnerships.

38. **ROI reports**: Assist in writing ROI reports for marketing campaigns.

39. **Webinar summaries**: Write detailed summaries of webinars for those who couldn't attend.

40. **Thank you notes**: Write professional thank-you notes for clients, partners, or event attendees.

41. **E-learning content**: Develop content for e-learning platforms.

42. **Franchise agreements**: Assist in writing franchise agreements for business expansion.

43. **Drip campaign content**: Write engaging content for drip email marketing campaigns.

44. **Scripts for interactive voice response (IVR) systems**: Create scripts for automated phone systems.

45. **Fundraising letters**: Write persuasive fundraising letters for charitable causes or projects.

46. **Earnings report summaries**: Write summaries of earnings reports for stakeholders.

47. **Greeting cards**: Write text for business greeting cards, such as holiday or anniversary cards.

48. **Invitation letters**: Write professional invitation letters for business events.

49. **Product comparison guides**: Create detailed product comparison guides for customers.

50. **Quizzes**: Develop quizzes for social media or email marketing to engage users.

51. **Sales presentation slides**: Write persuasive text for sales presentation slides.

52. **Sponsorship proposals**: Draft compelling sponsorship proposals for events or projects.

53. **Service level agreements**: Assist in drafting service level agreements (SLAs).

54. **Telemarketing scripts**: Write scripts for telemarketing efforts.

55. **Apology emails**: Write sincere apology emails for service disruptions or errors.

56. **Explainer video scripts**: Write concise scripts for product or service explainer videos.

57. **Meta descriptions**: Write SEO-optimized meta descriptions for website pages.

58. **Billboard ad copy**: Write short, impactful copy for billboard ads.

59. **Reactivation campaigns**: Draft emails for reactivation campaigns targeting dormant customers.

60. **Text for animated marketing videos**: Write engaging scripts for animated marketing videos.

61. **Market segmentation reports**: Assist in writing market segmentation reports.

62. **Radio ad scripts**: Write catchy scripts for radio ads.

63. **Data policy writing**: Help in drafting a comprehensive data policy.

64. **New feature announcements**: Write exciting announcements for new product features.

65. **Product recall notices**: Write professional product recall notices.

66. **Merger and acquisition announcements**: Write detailed announcements for business mergers or acquisitions.

67. **Customer profiling**: Assist in creating detailed customer profiles.

68. **Multilingual content**: Translate your content into multiple languages for international audiences.

69. **Loyalty program announcements**: Write engaging announcements for loyalty program launches or updates.

70. **Reminder emails**: Write friendly reminder emails for appointments, renewals, etc.

71. **Business card text**: Create catchy and memorable text for business cards.

72. **Holiday sales emails**: Write festive emails for holiday sales or promotions.

73. **Product update logs**: Write clear update logs for each new version of your product.

74. **Dispute responses**: Assist in drafting responses to customer disputes or complaints.

75. **Customer experience mapping**: Provide narratives for different stages in the customer journey.

76. **Out-of-office messages**: Write creative out-of-office messages.

77. **Outreach for guest speakers**: Write outreach emails to potential guest speakers for events or webinars.

78. **Event sponsorship emails**: Draft emails seeking sponsorship for events.

79. **Follow-up emails for networking**: Write follow-up emails after networking events.

80. **Employee survey creation**: Create detailed employee surveys to improve your workplace.

81. **Warranty information**: Write clear and detailed warranty information for your products.

82. **Product return policies**: Write easy-to-understand product return policies.

83. **SMS marketing messages**: Write engaging SMS messages for marketing campaigns.

84. **Strategic plan writing**: Assist in writing a strategic plan for your business.

85. **Voicemail scripts**: Write professional voicemail messages for your business.

86. **Training program content**: Develop content for internal or customer-focused training programs.

87. **Personalization of customer journey**: Help personalize content at each stage of the customer journey.

88. **Cancellation policy**: Draft a clear and fair cancellation policy.

89. **Compilation of industry news**: Compile and summarize important industry news.

90. **LinkedIn summaries**: Write professional LinkedIn summaries for your team.

91. **Startup name ideas**: Generate unique names for a new startup.

92. **Mission statements**: Help craft a powerful mission statement for your business.

93. **Loss prevention guide**: Write a guide to help customers prevent common mistakes or losses.

94. **Instructions for product assembly/use**: Write clear instructions for product assembly or use.

95. **TikTok captions**: Write catchy captions for your business's TikTok videos.

96. **Salary negotiation emails**: Assist in writing emails for salary negotiations.

97. **WhatsApp business messages**: Write professional messages to individuals and groups on WhatsApp Business.

98. **Content for augmented reality (AR) experiences**: Create content for AR experiences.

99. **Rebranding announcement**: Write a detailed announcement for a business rebranding.

100. **Exit interview questions**: Create a list of comprehensive exit interview questions.

101. **Email drafting**: Use ChatGPT to help draft emails, with a specific focus on marketing copy or business communication.

102. **Content creation**: Create blog posts, articles, or other types of content for content marketing strategies.

103. **Social media management**: Automate some aspects of social media post creation and response management.

104. **Ad copy creation**: Use ChatGPT to draft and iterate on advertising copy for various platforms.

105. **Website copywriting**: Improve your website's text to make it more engaging, informative, and persuasive.

106. **Market research**: Use ChatGPT to analyze market data and provide summaries or insights.

107. **Customer service**: Deploy as a chatbot for handling common customer inquiries.

108. **Product descriptions**: Write engaging and detailed product descriptions for eCommerce websites.

109. **SEO strategy**: Get suggestions for keyword optimization and SEO-focused content.

110. **Press release writing**: Draft press releases about company news or product launches.

111. **Newsletter writing**: Create engaging newsletters for email marketing campaigns.

112. **Brand storytelling**: Help craft your brand's story to connect with customers on a deeper level.

113. **Competitor analysis**: Use it to compile information from various sources about your competition.

114. **Idea generation**: Get new marketing ideas or tactics based on industry trends and data.

115. **Survey creation**: Formulate effective questions for customer or market surveys.

116. **Training materials**: Create educational content for training new employees or customers.

117. **Proposal writing**: Draft proposals for potential clients or projects.

118. **Translation assistance**: Translate basic messages or content for international marketing efforts.

119. **Brainstorming sessions**: Provide a unique perspective during brainstorming sessions for campaign ideas.

120. **Sales scripts**: Draft persuasive sales scripts for your sales team.

121. **Event planning**: Assist in planning and organizing business or marketing events.

122. **Recruitment posts**: Write engaging job descriptions to attract the best talent.

123. **Product naming**: Generate creative names for new products or services.

124. **Direct mail campaigns**: Help craft compelling copy for direct mail marketing efforts.

125. **Crisis communication**: Aid in formulating responses during a company crisis or PR issue.

126. **Slogan creation**: Invent catchy and memorable slogans or taglines for your brand.

127. **A/B testing**: Generate variations for A/B testing in marketing campaigns.

128. **Customer feedback analysis**: Help categorize and interpret customer feedback.

129. **Report writing**: Assist in writing business or marketing reports.

130. **Onboarding emails**: Create engaging welcome emails for new customers or subscribers.

131. **Hashtag recommendation**: Provide relevant hashtag suggestions for social media posts.

132. **Influencer outreach**: Craft persuasive messages for reaching out to influencers or brand advocates.

133. **Product launch plans**: Aid in writing comprehensive product launch plans.

134. **Interactive storytelling**: Use in interactive campaigns to engage with customers on a deeper level.

135. **Presentation preparation**: Assist in preparing presentations for pitches or business meetings.

136. **Content repurposing**: Help in repurposing existing content into different formats.

137. **Speech writing**: Draft speeches for company events or public addresses.

138. **Email subject lines**: Create compelling subject lines to increase email open rates.

139. **Customer testimonials**: Write customer testimonials based on their feedback.

140. **Chat support training**: Use its responses to train human chat support teams.

141. **Company FAQs**: Compile and write answers for a comprehensive FAQ section.

142. **Research summaries**: Provide summaries of market or business research.

143. **Internal communication**: Improve communication within the company by aiding in drafting internal memos or messages.

144. **Lead nurturing emails**: Create personalized emails for lead nurturing campaigns.

145. **UX writing**: Improve the user experience on your website or app with engaging and clear copy.

146. **Case studies**: Assist in writing compelling case studies to showcase success stories.

147. **Course creation**: Develop educational content for online courses or webinars.

148. **Public relations**: Craft persuasive pitches for journalists and media outlets.

149. **Market trends prediction**: Based on existing data, help predict upcoming market trends.

150. **Video scripts**: Write scripts for promotional or informational videos.

151. **Data visualization**: Assist in turning complex data into more digestible content or descriptions.

152. **Customer retention strategies**: Help draft strategies to increase customer loyalty and reduce churn.

153. **Content scheduling**: Assist in planning and scheduling content for different marketing channels.

154. **Personalized marketing messages**: Create customized marketing messages for different customer segments.

155. **White papers**: Write in-depth white papers on industry topics.

156. **User manuals**: Assist in creating detailed user manuals for your products.

157. **Social media profiles**: Write engaging bios and descriptions for your brand's social media profiles.

158. **Affiliate marketing**: Draft compelling pitches for potential affiliate partners.

159. **Brand guidelines**: Assist in writing comprehensive brand guidelines.

160. **Customer journey mapping**: Contribute to crafting narratives for different customer personas.

161. **Forum responses**: Assist in drafting responses to questions or comments on industry forums.

162. **Podcast scripts**: Write introductions, conclusions, or entire scripts for podcasts.

163. **Webinar scripts**: Prepare scripts for live webinars or online events.

164. **Post-event follow-ups**: Draft follow-up messages after business or marketing events.

165. **Meeting minutes**: Assist in creating accurate and detailed meeting minutes.

166. **Marketing plan writing**: Help formulate comprehensive marketing plans.

167. **Book summaries**: Create concise summaries of business or marketing books.

168. **Grant writing**: Assist in writing grant proposals for funding opportunities.

169. **Guest post outreach**: Draft persuasive outreach emails for guest blogging opportunities.

170. **Trend analysis**: Provide analysis or summaries of industry trends.

171. **eBook writing**: Help write comprehensive eBooks on relevant industry topics.

172. **Business plan writing**: Assist in formulating detailed business plans.

173. **Mobile app descriptions**: Write engaging descriptions for your mobile app on various platforms.

174. **Seminar scripts**: Assist in preparing detailed scripts for seminars or talks.

175. **Mission statement writing**: Help craft compelling mission and vision statements for your company.

176. **User-generated content prompts**: Create prompts to encourage user-generated content.

177. **Landing page copy**: Write persuasive landing page copy to improve conversion rates.

178. **Proofreading**: Use it to proofread written content for grammatical errors.

179. **Sales funnel optimization**: Assist in drafting content to improve your sales funnel.

180. **Exit-intent popups**: Craft persuasive messages for exit-intent popups to reduce website bounce rate.

181. **Caption writing**: Write engaging captions for social media posts or product images.

182. **Internal knowledge base**: Populate an internal knowledge base with informative articles and how-to guides.

183. **Customer re-engagement campaigns**: Draft personalized messages for re-engaging inactive customers.

184. **Crowdfunding campaigns**: Write compelling descriptions and updates for your crowdfunding campaigns.

185. **Response to reviews**: Craft professional responses to both positive and negative customer reviews.

186. **Sales page writing**: Create persuasive sales pages to boost your conversion rate.

187. **LinkedIn summaries**: Write professional LinkedIn summaries for company executives or the business itself.

188. **Customer satisfaction surveys**: Develop comprehensive surveys to measure customer satisfaction.

189. **Business blogging ideas**: Generate a list of relevant blog post ideas for your business blog.

190. **Legal disclaimers**: Assist in drafting legal disclaimers or terms of use for your products/services.

191. **Referral program copy**: Write engaging copy to promote your referral program.

192. **Pitch deck creation**: Assist in drafting a compelling pitch deck for investors.

193. **Job offer letters**: Write professional job offer letters to successful candidates.

194. **Membership emails**: Draft emails for membership renewals or exclusive member offers.

195. **Promotional material**: Write copy for promotional material like brochures, flyers, etc.

196. **Awards applications**: Assist in writing compelling applications for business or industry awards.

197. **CSR messages**: Craft messages that highlight your corporate social responsibility initiatives.

198. **Conference networking**: Draft messages for networking or collaboration opportunities at conferences.

199. **Webinar invitations**: Write persuasive invitations for your webinars or online events.

200. **Annual reports**: Assist in writing comprehensive annual reports.

HOW TO DEFINE AND INJECT YOUR BRAND VOICE WHILE USING CHATGPT

Maintaining the unique voice, tone and style of your business or brand is important especially in an AI-driven world and with so many tools and technologies around us that can help to power our businesses. It can be very challenging to ensure your voice remains throughout all your marketing and content initiatives when it's easy to copy and paste directly from ChatGPT. But preserving and enhancing your brand's unique identity helps you to stand out among the rest.

What Is a Brand Voice?

Injecting your business or brand's voice is a crucial aspect that embodies the values, vision, personality, style, and core principles of who you are and what you want customers to feel about your business or brand. Your brand voice is more than just the words you use; it is how you communicate with your audience and how you want your business to be perceived.

A powerful brand voice is one that can connect with your audience on a deeper level, allowing for greater engagement and brand loyalty. Whether your brand voice is authoritative, playful, intellectual, funny, or friendly can depend on various factors including your business type, audience demographic, target audience interests, industry, and culture.

When you define your brand voice, it helps to craft guidelines that detail the specific choices in language, tone, and sentiment that should be used in all communications. These guidelines help ensure all messaging, from social media posts to advertising copy to customer service interactions to print media, remain cohesive and consistent. Understanding the specific words that describe your brand can also be input into ChatGPT to get more accurate on-brand content.

More Importantly, when you integrate your brand voice into every aspect of your interactions with customers, including AI-based tools like ChatGPT, it ensures a seamless, authentic, and cohesive brand experience across all touchpoints. This helps to increase brand recognition and build trust and reliability along the way.

When you define your business or brand, you can train ChatGPT to echo your brand's tone, style, and language nuances, creating a harmonious blend of advanced AI technology and your brand's personal touch.

Here are some steps you can take to define and inject your brand voice into ChatGPT interactions.

1. **Language and Tone**: The language and word tone used in ChatGPT play a significant role in the information it puts out. If your brand is laid-back, informal, and friendly, you can direct ChatGPT to use a more conversational and casual tone in your content. If your brand is more technical and serious, you can instruct the platform to use a more formal language style. Remember, ChatGPT and other AI-driven platforms are machines that quite often will do what you tell it. You can program the AI to mirror the language and tone you desire during its interactions.

2. **Industry Jargon**: Depending on the industry, the use of specific jargon can be important for maintaining a formal or structured tone and style and can be critical to showing your expertise in the content output. For certain industries such as the legal or healthcare industries, the correct terminology adds a touch of authenticity and accuracy to the conversation. ChatGPT can

be instructed to use certain phrases or words that align with your industry so that it is shown in your content.

3. **Scripted Responses**: Every brand or business has or should have a way they respond to common concerns or questions from customers. Customizing scripted responses to frequently asked questions can help you to further define your brand voice and ensures consistent and accurate information is provided each time without you having to reinvent the wheel. The replies should not only be informative but also reflect your personality.

4. **Messaging Consistency**: Lack of consistency is one of the biggest mistakes I see business owners making especially when it comes to their content. To ensure ChatGPT spits out consistent answers to questions, regardless of the numerous ways a question can be asked, you have to start with consistency in your brand message and voice. If you're a fun and personable brand, don't feed the AI platform technical or very formal verbiage.

5. **Personalization**: Injecting personalization into interactions can help in brand building. If you are a brand that uses customer's first names or you start off the conversation with a phrase that is synonymous with your brand ("Hey there, everybody" or "Hey good people"), you want to inform ChatGPT of this type of personality and language.

6. **Emotions**: This is especially critical for customer service focused content as emotions are the deciding premise in which we understand if someone is happy with our brand or not. You can instruct ChatGPT to give you content as a response to recognizing unhappy, frustrated, or satisfied users. If you're using a chatbot on your website or in automated responses on social media, you can train it to identify specific keywords or phrases that may indicate frustration, happiness, or dissatisfaction with a customer.

7. **Taglines and Slogans**: If your company has catchphrases, slogans or certain unique terminologies that you're known for

using among your clients and in your branding and marketing, incorporate these into your prompts for ChatGPT. You want the content it produces to sound like you and your brand, not like a robot machine. Along with creating content quicker, you can enhance brand recognition.

8. **Error Message:** You likely have seen a website or two that has a compelling error page message when you reach a broken link that is aligned to their brand. You can use ChatGPT to do the same thing. Instruct it to provide a well-crafted error message that maintains brand voice and is creative or empathetic.

9. **Brand-Aligned Personas:** Consider developing a specific bot persona that aligns with your business or brand. This persona can include a name, voice and tone, language style, and a background story to humanize it. A bot persona should resonate with your audience and reflect the critical attributes of your brand especially when a human is not present such as business after-hours.

10. **Frequent Updates:** Update the content ChatGPT has produced on a regular basis to ensure it is up to date with product or service launches and changes in your company's branding or messaging strategy.

As AI continues to evolve with the massive amounts of new data that is produced every day, it's important to regularly review and audit your content and chatbot interactions to ensure consistent on-brand messaging and customer satisfaction.

Further, it is important to keep in mind that implementation is an iterative process. Monitor customer interactions with the AI, gather data and feedback, review the data you receive, and use the insights to improve your brand as you grow.

THE DIFFERENT TYPES OF ARTIFICIAL INTELLIGENCE

Artificial intelligence is a broad field that includes many different technologies. I believe it is important for all of us to learn about these ever-evolving technologies not just as they pertain to our lives and business but as they pertain to the world and how those around us might be using them. This book certainly be missing information without including some definitions around the different types of artificial intelligence out there.

Here is a brief breakdown of the different types of AI out there and some potential use cases and applications for businesses.

Rule-Based AI

This is one of the earliest forms of AI, where decisions are made based on a predefined set of rules. The system uses a series of "if-then" statements to come up with a final answer. This approach is useful for businesses in tasks where rules can be defined clearly, such as data entry and validation, workflow management, and simple customer service scenarios.

Machine Learning (ML)

Machine learning is a subset of AI where the system learns and improves from experience without being explicitly programmed. Machine learning uses statistical methods to enable machines to improve with experience. It can be used by businesses for tasks such as customer segmentation, automation, product recommendations, sales forecasting, and detecting fraudulent transactions.

Deep Learning (DL)

Deep learning is a subset of machine learning that uses artificial neural networks with several layers. These layers are deep structures of neural networks, hence the name – deep learning. These layers enable the machine to learn and make decisions on its own. Businesses can use deep learning for image and speech recognition, natural language processing, and customer sentiment analysis.

Natural Language Processing (NLP)

Natural language processing is a branch of AI that helps computers understand, interpret, and manipulate human language. It can be used in businesses for chatbots, voice assistants, sentiment analysis, text data summaries, and documentation process automation.

Computer Vision

Computer Vision is an interesting field of AI that enables computers to understand and label images. Applications of computer vision in businesses include quality inspection in manufacturing lines, facial recognition for security, and data extraction from images in documents.

Reinforcement Learning

Reinforcement learning is an area of machine learning where an agent learns to behave in an environment by performing certain actions and observing the results. Businesses can use reinforcement learning to optimize operational costs in areas such as logistics, inventory management, and energy usage.

Robotic Process Automation (RPA)

Robotic process automation is a technology that allows anyone today to configure computer software, or a "robot," to emulate and integrate the actions of a human interacting within digital systems to execute a business process. RPA can be used to automate repetitive, rule-based tasks that are usually done by humans.

For businesses, it's important to understand that AI is not a one-size-fits-all solution and AI-driven platforms and tools are the same. Different ones will work for different brand and businesses which is why it is important to understand the underlying premise of what AI is and then do your research to discover what really works best for you.

AI implementation should be done with clear goals and ethical considerations in mind. This includes understanding and mitigating potential bias in AI decisions, protecting customer data, and ensuring transparency in AI processes.

The types of AI mentioned above are just some of the most prominent and widely used categories, but the field of AI is constantly expanding and evolving as tons of data is created and put out into the world every day. There are other branches of AI, subsets of the types already mentioned, and hybrid approaches that combine AI along with various other methods. Some of these include:

Genetic Algorithms

These are inspired by the process of natural selection and genetics and are often used in search and optimization problems.

Fuzzy Logic Systems

Fuzzy logic systems is a form of many-valued logic which deals with reasoning that is approximate rather than fixed or exact. It's used in systems that need to reason in a "human-like" way, dealing with concepts that can't be expressed as true or false but rather as "partially true".

Swarm Intelligence

This is inspired by the behavior of groups such as ant colonies or flocks of birds. Swarm intelligence can be used for optimization and distributed problem-solving tasks.

Expert Systems

These are AI programs that simulate the knowledge and analytical skills of human experts. They're used in specific fields where they can provide advice or recommendations, such as medical diagnosis or stock trading.

Hybrid AI Models

These models combine different AI techniques to improve efficiency and performance. For example, neuro-fuzzy systems combine neural networks and fuzzy logic to handle uncertainty and a lack of precision.

Explainable AI (XAI)

As machine learning models become more complex, their decision-making processes can be difficult to understand. Explainable AI aims to make these models more understandable to humans, which is especially important in industries like healthcare, legal, and finance where transparency and trust are critical.

Transfer Learning

Transfer learning is a research problem in machine learning that focuses on storing knowledge gained while solving one problem and applying it to a different but related problem. This is an essential part of machine learning for tasks such as natural language processing where pre-trained models like BERT and GPT-3 are used.

Capsule Neural Networks (CapsNet)

Capsule neural networks is a new type of deep learning system introduced in 2017 which aims to overcome limitations of traditional convolutional neural networks (CNNs) for image recognition, including preserving hierarchical spatial relationships between features.

Federated Learning

This is a machine learning approach where the training process is distributed among many devices or servers holding local data samples, without exchanging them. This approach can be used to build more robust AI models without compromising data privacy.

Each of these AI methods have their strengths, weaknesses, complexities, and complications. Remember, AI is a continuous work in progress, and knowing the right one to use depends on the problem to be solved, available data, and specific requirements of the task. Depending on what business you have, it can be beneficial to work with an AI professional or consultant to determine the best approach for a particular situation.

WHAT IS GENERATIVE AI?

Generative AI is one of the most exciting advancements in artificial intelligence technology in recent years. This type of AI is capable of creating something new from existing patterns or data.

From composing music, drafting text copy, to generating realistic images, Generative AI offers an array of potential applications for businesses. Its operation often revolves around complex machine learning models such as Generative Adversarial Networks (GANs), Variational Autoencoders (VAEs), and transformers like GPT-3 for natural language generation.

How can generative AI be part of our everyday business lives? The truth is that it already is. But in case you didn't know, here are some simple applications that involve generative AI.

Content Creation

One of the most promising applications of Generative AI is in content creation. It has the potential to transform how brands and businesses approach content marketing strategies and content creation as a whole.

For instance, Generative AI can draft blog posts, articles, and social media content. A business that traditionally might have spent hours crafting a blog post can utilize generative AI to create a draft and then have a human editor refine the draft into publishable content. This drastically reduces the time and effort required for content creation, freeing up resources to focus on other areas.

Generative AI can also be used to create personalized content which ranges from personalized emails to custom news articles tailored to individual users' interests. Personalization is key in today's digital landscape as everyone wants to be spoken to in a unique way. With Generative AI, businesses can take personalization to a whole new level.

Design and Visualization

Generative AI also has applications in the design field. Generative adversarial networks (GANs), for example, can generate new images or modify existing ones. They can create original graphic designs, logos, or product mockups. Similarly, they can be used in website design, generating different layouts and designs based on user preferences.

For e-commerce businesses specifically, GANs can be used to create virtual models of products or generate additional imagery for existing products. In the real estate sector, Generative AI could be employed to create 3D renderings of properties, offering potential buyers a more immersive experience.

Product Development

When it comes to product development, Generative AI can be a game-changer. In the fashion industry, for instance, GANs can generate designs for new clothing items based on the latest trends. In the tech industry, Generative AI can create new software code, identify bugs, or optimize existing code.

In the field of pharmaceuticals and healthcare, Generative AI is being used to help design new drugs, cutting research time and making the process more cost effective. By understanding the structure and characteristics of existing drugs and diseases, AI can generate possible molecular structures for new medicines, dramatically reducing the time and cost of drug discovery.

Customer Interaction

Generative AI can power chatbots and virtual assistants, leading to more engaging and human-like interactions with customers. These AI-powered chatbots can understand user queries and generate responses that can accurately address their concerns, providing 24/7 customer service and freeing up human resources.

Risk Management and Security

Generative AI can also be used in risk management and security. For instance, it can be used to simulate different risk scenarios and generate strategies for dealing with them. It can also be used in cybersecurity, generating new encryption algorithms or creating adaptive systems that can respond to threats in real time.

Generative AI offers immense potential for businesses across various sectors when it is used transparently, ethically, and respectfully. Despite the challenges, the benefits Generative AI brings to the table are undeniable. Businesses that invest in understanding and leveraging this technology will likely gain a competitive edge, drive innovation, and optimize their operations. With the rapid pace of AI advancement, it's only a matter of time before Generative AI becomes an integral part of all of our toolkits.

Generative AI for Business Owners

If you're an entrepreneur or business owner and you're researching how and when to implement generative AI, here are some things you should consider:

- **Ethics and Responsibility**

  Generative AI, like any technology, can be misused. It is our responsibility to ensure we use this technology ethically, being aware of issues around generating misleading information or deepfakes, infringing on copyright by using generated designs or content that is similar to another, and protecting user data.

- **Quality Control**

While Generative AI can produce impressive results, it's essential to have humans in the loop to ensure the quality of output. AI models can sometimes produce outputs that are incorrect or lack common sense, so having human oversight is crucial.

- **Learning Curve**

  Be ready to learn. And AI is worth learning. Implementing Generative AI can be a complex process. While there are many pre-built models and tools available, integrating these into existing workflows and tailoring them to specific use cases can be a technical challenge that requires an investment in time and resources.

- **Ongoing Regulation**

  As AI technology continues to evolve, so does the regulatory landscape. As business owners and entrepreneurs, it is critical for us to stay up to date with the current and new laws and regulations to avoid potential legal challenges

Generative AI for Marketing Professionals

For marketing professionals specifically, Generative AI presents exciting opportunities as well as a few unique considerations:

- **Personalization at Scale**

  Generative AI can help create personalized marketing experiences, messages, and content for individual customers at scale which leads to better engagement and improved customer satisfaction.

- **A/B Testing**

  Marketers can use Generative AI to produce a variety of content for the purposes of A/B testing. This allows for more robust testing and optimization of marketing strategies and less time spent on what is not working.

- **Trend Analysis and Prediction**

By training on vast amounts of data, Generative AI can spot trends and make predictions about what kind of content or design will resonate with your target audience.

- **Visual Content Generation**

  From creating marketing collateral to product images and social media posts, Generative AI can help automate and diversify visual content production.

- **Ethical Implications**
- Marketers must ensure they're transparent about the use of AI in content generation as well. Deceptive practices could lead to consumer mistrust and legal issues. For example, audiences should know when they are interacting with AI, like chatbots, rather than human customer service representatives.

- **Brand Voice**

  Maintaining a consistent brand voice can be a challenge when using any AI tool for content generation. Marketers need to carefully train and monitor their AI tools to ensure they reflect the brand accurately and consistently. Before you can train though, you'll want to make sure that your brand voice and guidelines are in place and used throughout your company.

While Generative AI offers many benefits to businesses and marketing professionals, it's crucial to understand the basics and approach its implementation thoughtfully and responsibly. With the right strategy and considerations, Generative AI can be a powerful tool as our businesses and brands grow.

APPLICATIONS OF CHATGPT

As we know by now, ChatGPT is capable of generating human-like text based on the input it receives, making it an ideal solution for many use cases including customer service, content creation, research, training and education, and many other things.

Customer Service and Support

One of the most popular applications of ChatGPT is in customer service and support. By deploying ChatGPT as a customer service chatbot, businesses can provide 24/7 support to customers or visitors. These chatbots can handle a wide range of queries, from simple questions about a product to troubleshooting issues. This reduces the response time and frees up humans to focus on complex issues that require personality and nuance.

An e-commerce business can use ChatGPT to answer frequently asked questions about shipping, returns, refunds, inventory, and product details. If a customer wants to know the status of an order, they can simply interact with the chatbot, which will pull up their order details and provide a quick update.

Content Generation

ChatGPT is a very powerful tool for content creation. Whether you need it to draft blog posts, create social media captions, do keyword research for SEO, or develop product descriptions for Amazon.com, ChatGPT can do it quickly and efficiently. In addition, if you're stumped for time or have hit writer's block, it can help you to brainstorm ideas for new content. This helps to streamline the content creation process and reduces manual labor in terms of time and mental capacity.

A digital marketing agency can use ChatGPT to generate a first draft of a blog post or advertising copy, which can then be refined by a human writer. This saves everyone time and allows the human writer to focus on adding personality and creativity along with ensuring the content aligns with the brand's voice.

Education and Training

ChatGPT can be used to create interactive educational content, presentations for group learning experiences, answer students' questions, and provide personalized learning experiences. A language learning company can use ChatGPT to create dialogues in different languages for students to practice with.

In addition, businesses can use ChatGPT for employee training. The model can simulate real-world scenarios and guide employees through them all while providing instant feedback. This interactive and personalized approach to training can significantly enhance the learning experience in classrooms and companies.

Personal Assistants

ChatGPT can be used to create AI-powered personal assistants. These virtual assistants can help schedule appointments, send appointment reminders, draft emails, create lists or itineraries, and provide recommendations for local restaurants, movies, and places to see based on personal preference. By automating these routine tasks, ChatGPT powered personal assistants can save you time and help you to stay organized.

Research and Development

ChatGPT can also be used to analyze and summarize large volumes of data within a matter of seconds. This can be especially useful in fields such as academic research or market research where handling vast amounts of data is a regular part of the job.

Beyond these distinct areas, ChatGPT can be used in some other potential use cases as well that may require more attention to detail and factual, accurate information.

- **Mental Health Support**: ChatGPT can be used as a preliminary support system for people dealing with stress, anxiety, and other mental health issues. It absolutely is not a replacement for professional help, but it can provide comfort, words of motivation or encouragement, and guide users to appropriate resources for further assistance.

- **Entertainment and Gaming**: In video games, ChatGPT can be used to create dialogues for characters, generate immersive narratives, and even serve as an intelligent NPC (Non-Player Character) that can interact with players in real-time.

- **Medical Assistance**: ChatGPT can be used to help people understand complex medical jargon by providing simplified definitions and explanations. It can also help answer general health-related queries. Similar to mental health support, it should never replace a human doctor or professional medical advice.

- **Coding Assistant**: ChatGPT can be trained to help developers by creating code and can be trained to understand programming languages as well as assist in debugging.

- **Simulation and Training**: ChatGPT can also be used to simulate conversations in a controlled environment. This is particular useful when role play may be needed for learning such as in sales, debates, negotiation, and customer support.

The applications and use cases of ChatGPT are vast and can be used in almost every industry. But its outputs should be monitored and moderated to ensure accuracy and appropriateness. The output from ChatGPT must be supervised, reviewed and edited, especially when used in sensitive areas. As with any AI technology, it's important to strike a balance between automation and injecting human personality.

10 STEPS TO TAKE TO ENSURE AI IS USED ETHICALLY IN YOUR BUSINESS

As we venture deeper into the digital era, artificial intelligence is becoming an integral part of our businesses, driving unprecedented innovation and growth. But this powerful tool comes with an even greater responsibility: to use it ethically. In this chapter, I will share ten steps you can take as a business owner ensure AI is used ethically in your business.

Navigating the complexities of AI ethics can seem daunting, but it is essential for long-term sustainability and for the reputation of your enterprise. From understanding biases to respecting privacy, from transparency in decision-making to accountability for AI based actions – these issues can impact your brand reputation, customer loyalty, and the overall trust of your business.

Each of the ten steps will provide detailed insights and practical tips to help you navigate the nuances of ethics in AI and offer you practical action steps to think about with your team members. The journey to ethical AI isn't simple, but it's critical to understand as we progress in this digital age.

Here are ten steps you can take to jumpstart your journey in the ethics of AI.

1. **Educate yourself and your team**: A lot of times, we make mistakes in our business because we don't take the time to learn. So, take the time to learn about the implications of AI, its benefits, limitations, and ethical challenges. Encourage your

team to do the same or take it up as a group project. Consider arranging workshops or training sessions with experts in the field to help take your team's understanding to the next level.

2. **Develop an AI ethics policy**: Work with stakeholders across your organization and even outside consultants and experts to create a policy that addresses how you will use AI and how you will handle customers' data. If you're in marketing, this is especially important when it comes to email marketing and advertising data. The policy you create should reflect your organization's values and ethics, as well as industry best practices and relevant laws and regulations.

3. **Transparency in AI interactions**: Make it clear to your customers when they're interacting with AI such as a chatbot on your website, or automations set up to respond in your DMs. In some cases, you will also want to make it clear when you're using content from an AI powered tool like ChatGPT. If AI is involved in making decisions that impact them, such as refusing a loan application based on specific parameters or making a high-risk investment, explain how those decisions are being made. As AI takes up a greater space in our lives, it is important to be transparent about what decisions and information is being shared with the AI and who is responsible if something goes wrong with its outputs.

4. **Prioritize data privacy**: Be transparent about how your company collects, uses, stores, shares, and protects data. In some cases, this is in the form of a Privacy Policy or a Terms of Use document on your website or mobile application. But these types of documents should pass through the eyes of a lawyer first before being posted. When in doubt, please ask an expert. Only collect data that you need and obtain explicit consent from users before collecting and/or sharing that data with anyone else.

5. **Bias testing is important**: AI bias is when an algorithm creates a result that is prejudiced because of error-laden assumptions in the artificial intelligence algorithmic process. To avoid this

type of issue, regularly test your AI systems and technologies to make sure they're not perpetuating or amplifying biases. When you use AI tools and platforms, any errors is ultimately a reflection of that tool or platform, but your customer doesn't know that and will place the blame with you. Testing could involve using diverse datasets or using techniques such as adversarial testing to try to identify biases in the behavior of your AI.

6. **Bring in the humans**: A human or several humans should always oversee and ultimately give the green light on any decisions made by an AI. If you're using ChatGPT to produce content at a large scale for your business, a human reviewer or editor should look at it and give it the okay to be published or distributed. This is particularly important for decisions that could have serious implications such as legal, health, and finance decisions for individuals or for your business.

7. **Protect yourself**: Invest in the security of your AI systems. Just like any other platform or tool, AI systems can be hacked into and as a result, the outputs it gives will be erroneous. Protect your AI systems from attacks by investing in robust security measures. This could involve things such as secure coding practices, regular security audits, and training for your team on recognizing and preventing security threats. Just because it is AI, and it is trained on all these models and datasets does not mean that it is exempt from failing at times.

8. **Retrain and reskill where necessary**: If we're being honest, AI is going to absolutely replace some jobs simply because the financials and the people management systems won't make sense for some companies going forward. Rather than striking off your team members at a moment's notice, take the time to offer them retraining and reskilling opportunities so they can learn a higher level skill and keep or transition into a job that really needs that level of expertise. If you're running an agency, you may not need entry-level or junior writers because AI can handle these tasks. But what you will need is someone to think

critically and be able to manage the AI along with people who are highly specialized and trained in understanding highly regulated industries and subject matters.

9. **Keep up to date with AI legislation**: AI legislation is coming quickly and is continually evolving, so it's important to stay informed about these changes that could ultimately impact your business if you rely on AI on a daily basis to accomplish your business goals. For example, in October of 2022, President Joe Biden released an AI Bill of Rights which essentially outlines the five protections that Americans should expect to have in light of AI. The European Union is finalizing their version of what rules they will enforce with regards to artificial intelligence. Almost every state in the United States has some level of legislation on the state side that is pending, being reviewed, or almost finalized. It is important to stay up to date on the state, national, federal and international legislation so you can make changes when and where needed.

10. **Get expert advice**: Consulting with experts in AI ethics, law, and security can be very beneficial to your business. Whether you employ a specific document or agreement from them or not, getting the knowledge and understanding from an expert in the field can help you make informed decisions as you move forward. These consultants can provide valuable insights and advice that are specific to your situation and needs but do your own due diligence to ensure you're speaking to the right person and that they know what you need them to know.

Ethics in AI is an ongoing conversation rather than several boxes that are checked on a long list of to-dos. The ethical landscape of AI is continually evolving, and as a business owner or marketing leader, it is important for us to stay updated and engaged with the latest discussions and best practices. Businesses that prioritize ethical considerations when deploying AI are more likely to build trust with their customers and stakeholders while safeguarding their long-term success.

WHAT IS PROMPT ENGINEERING?

Prompt engineering is a technique used in the context of language models like ChatGPT, where the goal is to curate an input prompt that will help to guide the AI to produce the desired output. In essence, it's about asking the right type of question or inputting the right type of text to get the most useful and accurate response.

This is where many people fail to use ChatGPT and other AI tools like it for content creation in the right way. And this is also where the machine still requires humans to think and decide what type of information is most useful for input so that it will output the right type of result.

Prompt engineering is based on the idea that language models learn to respond to inputs by understanding patterns in the data they have been trained on. The prompts serve as guidance, helping the AI focus on generating a certain type of response. By carefully creating these prompts, we can steer the model's responses towards specific topics, themes, styles, and formats that work best for our business.

At the end of this book, there is a list of actual prompts you can copy and use with ChatGPT specific to various scenarios that might occur in your business. As in most things, this platform will work well if you work it well.

There are several strategies we should consider using when it comes to effective prompt engineering:

- **Explicit Instructions**: Be clear and specific with the prompt you use. For example, instead of saying, "Translate this to Spanish," say "Translate the following English text to Spanish." And paste the actual text you want to translate below the prompt.

- **Request for Reasoning**: Ask it to provide reasoning for the answers it is giving you. This often helps it to obtain more detailed and reliable responses.

- **Role Setting:** You can guide the model to behave in certain ways by giving it a specific role to play. For example, you can instruct the model to behave like "a personal assistant that speaks like Shakespeare" and it will produce responses in a Shakespearean style.

- **Sample Outputs**: If you provide an example of the desired output along with the input prompt, it can help guide the AI model towards the type of answer you want it to produce. For example, you can tell it to write in the style and tone of your favorite writer and it will do that to the best of its ability.

Prompt engineering is both an art and a science and requires an understanding of how the AI model interprets and responds to different types of prompts. However, it is not difficult to do. With some practice and testing, you can also come up with a set of prompts that are specific to your business and by which you can train your team to use when needed. Knowing what prompts will give you the desired output is a crucial skill and is part of making the most out of AI language models in various applications, including content generation, data analysis, market research, and customer service.

Let's dive a little bit deeper into some advanced prompt engineering techniques:

- **Injecting Constraints**: To get a specific type of output, you can inject constraints in your prompt. For example, asking "Explain quantum physics in simple terms that a 10-year-old will understand" helps to injects the constraint of simplifying complex concepts to the level of a young child. This is definitely an interested leg up that ChatGPT has over Google Search. Rather than sifting through multiple entries to find the most basic definition and understanding, ChatGPT can craft something that is exactly what you need.

- **Sequential Prompts**: In a conversation with ChatGPT, every new prompt includes the entire history of the prior conversation which means the AI considers all previous exchanges in generating its responses. By revisiting a request or building on previous prompts, you can guide the conversation and get the most accurate results as possible. Building on the prompt example above, you can ask it to, "Give me a 500 word essay explaining quantum physics in simple terms that a 10-year-old will understand." And it will take what it gave you the first time and revise it into a lengthier piece of content based on this second prompt.

- **Prompt Chains**: You can break down a complex task into smaller tasks and feed these smaller tasks to the AI model one after the other. For instance, if you want to summarize a scientific paper, you could first ask the model to explain the main concepts in the paper, write a 150-word summary, condense the main concepts into a 300-word essay, and write 10 test questions with answers based on the content.

- **Prompt Programming:** This is great for engineers and developers. It involves writing code-like instructions in natural language in the prompts. If you want to cut your development or coding time in half, you can instruct the AI model to perform complex tasks with a series of steps in which the output will be generated in code.

With a little time, effort, understanding, and testing, anyone can learn how prompt engineering works and how to create prompts that get you the results you want.

It's a continual process of iteration and refinement: you propose a prompt, evaluate the output, refine the prompt based on the results, and repeat the process.

As you gain more experience and intuition about how the model responds to different types of prompts, you'll be able to more effectively use prompt engineering to get the results you want and create prompts for various aspects of your business including weekly content, website coding, customer service problem resolution, and a host of other things.

WHAT IS A CHATGPT PROMPT?

I n the previous chapter, we talked about prompt engineering, what it is, and how to use it within the context of ChatGPT. In this chapter, we are going to discuss what a ChatGPT prompt looks like and how to write a prompt to get the best output.

If you are using ChatGPT to write presentations, blog posts, meta descriptions, website content, email marketing campaigns, landing page copy, social media captions, and a host of other things for your business, it is important to make sure you are feeding it the right information about your business, how you speak about your business, and how you want customers to feel about your business so that it gives you exactly what you need.

So, what is a ChatGPT prompt?

A ChatGPT prompt is the text that you provide or input into the AI model to request a response, answer a question, generate a command, offer a greeting, or strike up a conversation. The ChatGPT model generates a response based on this input or prompt you provide it. This is why it is very important to know how to create prompts so that you get the right output from ChatGPT.

Here are a few examples of prompts you could give to ChatGPT:

- **Greeting:** "Hello! How are you today?"

- **Direct Question:** "What is the weather like in New York City?"

- **Command:** "Tell me a joke."

- **Instructive Task:** "Translate the following sentence to French: 'The quick brown fox jumps over the lazy dog.'"

- **Ask a Question:** "Can you explain the theory of relativity in simple terms?"

The response from ChatGPT always depends on the nature of the prompt. For example, if you ask a question, the model will attempt to provide an answer based on the data it has been trained on. If you provide a command or task, the model will attempt to carry it out to the best of its ability.

How to Write Prompts to Get Best Results

Crafting an effective prompt for ChatGPT is both an art and a science. Here are some general guidelines and examples that can help you write effective prompts for your brand or business:

1. **Give Specifics**: The more specific and detailed your prompt is, the better the AI can generate a useful response. If you're looking for an answer to a question, try to make your question as clear as possible.

    o Less Effective: "Tell me about batteries."

    o More Effective: "Explain how a lithium-ion battery works."

2. **Set the Context**: If your task involves a specific context, make sure to include it in the prompt. This can guide the AI model to provide a response that specifically fits your needs.

    o Less Effective: "Write a product description."

    o More Effective: "Write a product description for Amazon.com about a lightweight, portable camping tent that sleeps six people and is weather-resistant."

3. **Request a Specific Format or Word Count:** If you need the answer or output to be in a particular format, style, or with a specific word count, add this to your prompt instructions.

    o Less Effective: "Tell me about content marketing."

o   More Effective: "Give me a 700 word blog post about how to craft a content marketing plan for my local bakery."

4. **Consider Prompt Chains**: If you have a complex task, you can break it down into a series of smaller prompts. This allows the AI to focus on one aspect of the task at a time. In this case, you receive an output based on each individual question.

o   First Prompt: "Who was Albert Einstein?"

o   Second Prompt: "What was his contribution to physics?"

o   Third Prompt: "Can you explain the theory of relativity in simple terms?"

5. **Iterate and Refine**: Sometimes, you may not get the desired response on your first attempt. All you need to do is refine the prompt or instructions you're giving the AI. Modify your prompt or ask the question in a different way to get the right type of information you need.

o   Initial Prompt: "How do I make a paper airplane?"

o   Refined Prompt: "Can you provide detailed, step-by-step instructions for how to make a paper airplane?"

6. **Infuse Creativity**: For creative tasks, you can ask the AI to take on a certain role or write in a specific style using words that define what level of creativity you would like.

o   Less Effective: "Write a story about a dragon."

o   More Effective: "Write a short children's story about a friendly dragon who loves to bake cookies and deliver them to sick children."

o   Most Effective: "Write a short children's story of 350 words about a friendly dragon who loves to bake cookies and deliver them to sick children. Set the story in the mid 1940s in New York City at the hospital in Manhattan."

Getting great results from ChatGPT begins with giving the AI model great prompts. It may involve some level of trial and error as well but the more you experiment with it, the better you will become.

HOW TO DEVELOP A CONTENT STRATEGY USING CHATGPT (WITH EXAMPLES)

In this digital age, compelling content is at the heart of successful online engagement. There is no business that runs well without some form of effective content strategy in place. However, creating a robust content strategy that resonates with your audience can often seem like a Herculean task. But it doesn't have to be this way.

ChatGPT can augment your content strategy in ways that may have been unimaginable for one person or a small team to do on their own. As a sophisticated AI language model, it can create content that is not only relevant and engaging but also personalized to your target audience's preferences and interests.

In this chapter, we will discuss the process of integrating ChatGPT into your content strategy. From understanding the basics of ChatGPT and its capabilities as has been described in previous chapters, to practical tips on leveraging it for different types of content – blog posts, social media captions, email newsletters, and advertising copy.

A good content strategy is key to engaging your target audience and achieving your overall business goals. Here are some eight key steps to consider:

1. Define Your Goal

What do you want to achieve with your content? Goals could range from increasing brand awareness, driving traffic to your website, generating leads, boosting sales, or improving customer retention. Once you have a goal defined, you can use ChatGPT to help you develop content that fits within the various stages of achieving that goal.

2. Define Your Audience

Understand who your target audience is who will receive the content you produce. Who do you want to reach? What are their interests, needs, and pain points? What types of content do they consume and where do they spend most of their time online? This understanding will help you create content that really resonates with the people you want to reach. This is where you will want to integrate Google Analytics, Search Console, Google Trends, Buzzsumo, and insights from your social media analytics to learn more.

3. Conduct a Content Audit

If you've been creating content for some time already, now is a good time to evaluate what is working, what isn't working, and what you need to add or start testing in your process. A content audit is an ongoing part of your content strategy. Which pieces of content have been most successful in terms of engagement, shares, or conversions? Which haven't performed as well? Are you focused on the right platforms? Use these insights to guide your content creation process.

In short, your content audit process should include at a minimum these four steps:

- **Create an inventory of all the content you have created**: It is useful to create a spreadsheet at this phase. Useful information to include is URL, title, content type, author, and publication date.

- **Define the quantitative metrics to analyze your content:** These metrics can include all or any combination of traffic to the page, bounce rate from page, conversion rates if conversion goals were set, social shares such as likes, engagement, mentions and comments, as well as SEO rankings such as positioning in SERPs.

- **Conduct a qualitative review against specific factors:** These factors can include all or any combination of whether the content aligns to the brand, if it's still relevant to the audience, and whether the user experience is correct.

- **Gather and note the insights, patterns, adjustments, or recommendations you would make based on the outcome of the above factors:** Based on the quantitative and qualitative outcomes above, you can see what is performing/not performing, what topics resonate/or don't resonate with your audience and missed opportunities or gaps in your content. These can all be used as part of your content strategy growth plan.

4. Identify Content Types and Platforms

Based on your content goals, audience analysis, and content audit, identify the types of content and platforms that work best for your audience. These could include blog posts, videos, podcasts, social media posts, infographics, newsletters, webinars, whitepapers, or eBooks. Also, think about the platforms where you'll distribute this content, such as your website, email list, text message lists, Facebook, Twitter, Instagram, LinkedIn, Threads, YouTube, through a podcast, or through collaborations with peers in your industry.

5. Decide Who and How Content Will Be Created and Managed

In this step, you will want to decide two key things: Who will create what type of content and how will the content be managed so everyone is on the same page? At my agency, DW Creative Consulting Agency, we manage content and web projects for a range of companies and brands all through Monday.com. I'm not getting paid to promote them or include them here, but they are my personal favorite for the entire lifecycle of project management. But there are others that you can research and use as well.

Defining roles and responsibilities is crucial at this stage and it is okay if you are a team of one. If you are creating content solo, you need to know your bandwidth and what you can realistically do consistently. If you're creating content with others, you need to know each person's bandwidth and where they fit into the content creation process.

6. Create a Content Calendar

A content calendar helps you plan, organize, and schedule your content. It outlines what type of content will be created, when it will be published, where it will be shared, and who is responsible for each part or the entirety of the process. Consistency is key in content marketing. It really is equally about quantity and quality when it comes to keeping up in this digital age.

The brand or business who does content the best and the fastest more than likely wins always. So, make sure your calendar is set for you and your team to regularly produce and publish content. ChatGPT is your friend in this matter. Unlike humans who need breaks and weekends and sick days; ChatGPT is always there to help you stick to a schedule.

7. SEO Strategy

The importance of SEO cannot be stated enough. You don't just create content for yourself or for the 300 people on your email list. If you are publishing content to your website or any other public forum, search should be a part of the strategy and process.

Optimize your content for search engines to help increase visibility. You can do this by identifying relevant keywords and using them strategically in your content. You can also do a competitor analysis and a content gap analysis to figure out what types of content topics and formats and other opportunities you may be missing out on. In addition, your SEO plan should include using meta descriptions, alt tags for images, fixing technical errors on your website, and creating backlinks.

You also don't want to get about YouTube SEO if you're heavy on video. YouTube is the second largest search engine (behind Google, of course). Research target keywords to include in your video content titles and descriptions so your ideal audience can find you.

8. Measure and Adjust

Peter Drucker said, "What gets measured gets managed." You can really manage anything that does not first get measured. Measuring your results looks like asking yourself: Am I meeting my goals? What types of content are most successful for me and my business? Track your progress and adjust your strategy, not your goals, as needed.

A content strategy is not a one-time, set it and forget it type of process. It should change and evolve just as much as your business changes and evolves over time. It should adjust with your audience's needs and interests. In short, a good content strategy should be fluid and adaptable where needed.

Example of a Content Strategy in Action for a Coaching Business

Let's get into this example of how a coaching business might approach content strategy:

1. **Define Your Goal**: As a coaching business, your goal might be to build your brand's reputation, attract new clients, and nurture existing ones. Specific goals could include increasing website traffic by 50% in the next six months or acquiring 20 new clients in the next quarter.

2. **Understand Your Audience**: Your target audience could be individuals looking for personal development, long-term wellness strategies, or career advancement. You must know their needs, pain points, interests, and aspirations. Are they looking for guidance, motivation, resources, or techniques?

What age group and professions are they in? Use surveys, existing client interviews, or market research to gather this information.

3. **Conduct a Content Audit**: Assess your existing content. Do your social media posts receive comments and shares? Do your how-to guides attract email newsletter sign-ups? Is there a topic or format that resonates more with your audience? Has your audience had a coach before or are they at the beginning stages of their journey? Use these insights to identify successful content types and themes.

4. **Identify Content Types and Platforms**: For a coaching business, content types might include blog posts (with tips, advice, and industry insights), webinars, podcasts, instructional videos, downloadable resources (like eBooks or checklists), and client testimonials. You might distribute this content through your website, email newsletters, LinkedIn (for professional coaching), Instagram or Facebook (for lifestyle or health coaching), and platforms like YouTube for videos.

5. **Create a Content Calendar**: Plan out when and where you'll publish your content. For instance, you might decide to publish a blog post every Tuesday, a brief summary of a topic on social media every morning, a detailed guide once a month, and host a webinar every quarter.

Content Calendar Table Example for Coaching Business

Here's an example of how you might structure a content calendar for one month in your coaching business:

Date	Content Type	Topic/Title	Channel	Status
January 1, 2023	Blog post	"5 Techniques for Stress Management"	Website, LinkedIn, Email newsletter	Drafting
January 4, 2023	Social media post	Motivational Quote	Instagram, Facebook	Scheduled
January 6, 2023	Podcast episode	"Effective Communication for Career Growth"	Website, Apple Podcasts, Spotify	Editing
January 8, 2023	Email newsletter	Monthly Tips & Resources Roundup	Email	Scheduled
January 11, 2023	Blog post	"Setting and Achieving Personal Goals"	Website, LinkedIn, Email newsletter	Ideation
January 13, 2023	Webinar	"Work-Life Balance Mastery"	Website, Email (invitations)	Planning
January 15, 2023	Social media post	Client Testimonial	LinkedIn, Facebook	Scheduled
January 18, 2023	Video	"Step-by-Step Guide to Mindfulness Meditation"	YouTube, Website, Social Media	Filming
January 21, 2023	Infographic	"Key Milestones in Personal Growth"	Instagram, Facebook, Pinterest	Designing
January 24, 2023	Blog post	"Overcoming Imposter Syndrome"	Website, LinkedIn, Email newsletter	Ideation

A content calendar can be a flexible tool, but it is designed to help you stay organized, on schedule, and consistent with your content creation and distribution plan.

6. **Decide Who Does What and When**: As a coaching business owner, you might create most of your content in-house since it relies on your personal expertise. However, tasks like content editing, graphic design, and social media management can be outsourced to others. Define who's responsible for what within your team and use a good project management system to keep up with deadlines.

7. **SEO Strategy**: Identify relevant keywords related to your coaching field (like "career coaching tips" or "wellness for a busy professional" or "healthy lifestyle advice") and incorporate them into your content. Using SEO-friendly practices will help your content rank higher in search results.

8. **Measure and Adjust**: Track your progress always. Are you gaining more traffic to your website? Are you converting this traffic into coaching clients? Do your webinar attendees sign up for your services? Are you getting more email subscribers? Are you tracking well against your financial goals? Use these insights to continuously improve your content strategy.

Having a content strategy ensures your content aligns with your business goals and your audience's needs and preferences. The key to a successful content strategy is consistent, quality content that provides continuous value to your audience. These eight steps will help you to get on the journey to crafting a strategy that works for you.

HOW TO INJECT CREATIVITY INTO AI-WRITTEN CONTENT

While AI is great for generating content efficiently, it can lack human personality and creativity at times. This is why I tell my clients that ChatGPT is a true companion to your content creation initiatives. It can help you get to where you need to go faster in your business. But you are ultimately responsible for injecting your brand voice, tone, style, words, and personality into what it outputs, so you have content that reflects you.

Here are some ways you can infuse creativity into AI-written content from ChatGPT and other AI writing tools:

1. Give Unique Inputs

AI systems work based on the input provided to them. If you provide an interesting and unique prompt, the AI is more likely to produce a creative output. Experiment with the different types of prompts you can give it to see which ones generate the most unique responses.

2. Blend AI and Human Writing

Consider using AI-written content as a starting point or as a first draft for a lot of your content initiatives. It can draft the basic structure and details, and then you or a writer you select can come in to add the personality, creativity, emotion, and uniqueness that the AI content may be missing. Don't get frustrated with the AI model not giving you absolute perfect content outputs especially if you're not giving the prompts your all.

3. Consider Using AI for Ideation

AI can be a powerful tool for brainstorming and generating new ideas. It can generate a multitude of ideas around a theme or topic, which you can then refine and expand on. These ideas can spark creativity for a new idea and often lead to unique content pieces. If you find prompt engineering too difficult, use AI models as a resource for ideas.

4. Content Formats Can Be Fun

AI isn't limited to just generating new blog text or listicles, especially if you're using the paid version of ChatGPT. It can also create content in other formats such as poetry, video scripts, podcast show notes, dialogue for fiction stories, and text to graphics. Exploring different formats can result in more creative content.

5. Customize the Output

Many AI systems allow users to customize the tone, style, voice, and other parameters of the output. Experiment with these settings to create a variety of content that stands out. One thing that helps here is having a word bank to describe your brand and injecting these words into the AI model, so it continuously produces content that stays on brand.

6. Curate and Refine

AI often generates more content than what is needed. A key to creativity can be curating the best parts from the AI-generated content and refining or combining them in novel ways.

7. Continual Learning and Feedback Loops

As AI systems continue to improve, they're increasingly able to learn from the feedback we give it. You will notice when you ask ChatGPT to "regenerate response," it does so and then will ask if the second or third version was better, worse, or the same as the versions before it. By providing the system with feedback on its outputs, over time, it can better understand the type of creative content you're seeking.

8. Be Human as Much as is Humanly Possible

While AI can generate massive amounts of content in a short period of time, please remember to be human as much as possible and keep a personal, relatable attitude for your audience sake. This is especially needed when responding to comments or direct messages from your audience. Personal, human responses add creativity and authenticity that should not be lost with AI.

While AI can serve as a powerful tool in content creation, it works best as a collaborator rather than a replacement for human creativity, personality, and intuition. By adding your creative touch and editing the AI-generated content, you can leverage the strengths of AI while maintaining the distinctiveness and uniqueness of your brand or business.

HOW TO USE CHATGPT TO BRAINSTORM IDEAS

C hatGPT is an excellent tool for brainstorming ideas. From new business names, baby names, travel itineraries, meal prep plans, dinner ideas, blog posts topics, and product names, ChatGPT has got you covered.

Here are seven things to consider when you're working with ChatGPT to brainstorm your next big idea.

1. **Clearly State Your Needs**: Be explicit about what you're looking for. If you need ideas for blog post titles or topics, for instance, specify the industry, audience, and style you're aiming for. A well-structured prompt might be "Generate 20 blog post titles for a vegan food blog targeting millennials interested in quick and easy recipes."

2. **Ask for Variations**: If you already have a basic idea but need to explore it further, ask ChatGPT for variations. For example, if you're naming a new productivity app, you could say something like this, "Generate 10 names for a productivity app focused on helping remote workers manage their time effectively."

3. **Explore Different Angles**: Let ChatGPT help you to consider different perspectives or angles on a topic. If you're planning a marketing strategy for a new product launch, you might ask, "What are different marketing strategies for launching a sustainable fashion brand?" or "What are the best ways to promote the release of my new line of sustainable footwear."

4. **Use Open-Ended Questions**: Open-ended prompts can yield more creative results. Asking something like "What are some innovative ways to use AI in education?" could spark an interesting brainstorming session with the AI model.

5. **Ask for Pros and Cons**: If you have an idea but aren't sure about its potential, you can ask ChatGPT to list the pros and cons for you. This can help you evaluate the idea's viability and identify potential challenges or benefits you may not have already considered. Don't forget to actually test your idea in a live market once it is ready.

6. **Iterate and Refine:** If the AI's first responses aren't quite what you're looking for, refine your prompt or ask follow-up questions to guide the AI in the direction you want it to go. Just asking it for "Ideas on [topic]" might not give you as focused of a view as you want.

7. **Combine Ideas:** Use ChatGPT's output as a starting point and combine or modify all of the ideas it has helped you come up with into something unique that works for you.

ChatGPT can be a powerful brainstorming and ideas generation tool. Take the ideas it generates, iterate on them, and apply your own creativity and critical thinking to develop them further.

Over 300 Examples of Ideas Prompts from ChatGPT

Next, here are some actual very broad examples of prompts generated from ChatGPT that you can use to brainstorm your next set of ideas. These examples cover a wide range of topics from product development to virtual reality and healthcare to pets.

Feel free to use this list as a starting point for your next big idea.

1. "Generate 10 tagline ideas for a new fitness app."

2. "What are some interesting blog post topics about vegan cooking?"

3. "Suggest 5 unique features for an online tutoring platform."

4. "What are some innovative uses for drones in agriculture?"

5. "Generate ideas for promoting a new book on social media."

6. "Suggest names for a new brand of organic skincare products."

7. "What are 5 ways we could improve our customer service?"

8. "Generate ideas for interactive exhibits for a science museum."

9. "What are some potential uses for VR in real estate?"

10. "What are some creative themes for a tech conference?"

11. "Generate 10 concepts for mobile games that teach kids math."

12. "What are some unique selling points for a zero-waste coffee shop?"

13. "Suggest names for a new line of sustainable activewear."

14. "What are some possible benefits of using AI in healthcare?"

15. "Generate ideas for weekly segments on a sports podcast."

16. "What are some ways we could make our office eco-friendlier?"

17. "Generate topics for a YouTube channel about personal finance."

18. "What are some innovative functionalities we could add to our travel app?"

19. "Suggest names for a new type of tropical houseplant."

20. "Generate ideas for engaging Instagram posts for a pet supply store."

21. "What are some potential applications for blockchain in education?"

22. "Generate taglines for a new brand of electric bikes."

23. "What are some creative ideas for a company team-building event?"

24. "Generate ideas for fundraising initiatives for a wildlife conservation charity."

25. "What are some ways we could increase diversity in our company?"

26. "Generate topics for an email newsletter about digital marketing."

27. "What are some unique features for a new dating app?"

28. "Suggest names for a new craft beer."

29. "Generate ideas for promoting a new album on TikTok."

30. "What are some potential uses for 3D printing in fashion?"

31. "Generate ideas for DIY tutorials for a home improvement blog."

32. "What are some strategies for improving employee wellness?"

33. "Generate taglines for a new cooking show about desserts."

34. "What are some ways we could reduce waste in our manufacturing process?"

35. "Generate ideas for special promotions for a bookstore during the holidays."

36. "What are some potential applications for AI in the event planning industry?"

37. "Generate 10 ideas for digital marketing campaigns for an online bookstore."

38. "What are some innovative features we could add to our fitness tracking app?"

39. "What are potential themes for our next corporate retreat?"

40. "Generate 20 ideas for engaging social media posts for our organic skincare brand."

41. "Generate themes for a photo contest on a photography website."

42. "What are some innovative designs for a new line of running shoes?"

43. "Suggest names for a new brand of eco-friendly cleaning products."

44. "Generate ideas for charity events for an animal rescue organization."

45. "What are some ways we could improve the user interface of our website?"

46. "Generate blog post topics for a website about sustainable living."

47. "What are some creative designs for a new line of kids' backpacks?"

48. "Suggest names for a new health food restaurant."

49. "Generate ideas for YouTube videos for a fitness influencer."

50. "What are some potential uses for AI in the music industry?"

51. "Generate ideas for social media campaigns for a women's clothing brand."

52. "What are some ways we could increase accessibility in our app?"

53. "Generate taglines for a new documentary about climate change."

54. "What are some innovative ideas for a new line of smart home products?"

55. "Suggest names for a new tech startup that uses AI for language translation."

56. "Generate ideas for podcast episode topics about entrepreneurship."

57. "What are some potential applications for virtual reality in therapy and counseling?"

58. "Generate ideas for blog posts for a website about hiking and outdoor adventures."

59. "What are some ways we could improve our customer loyalty program?"

60. "Generate ideas for Instagram Stories for a travel blogger."

61. "What are some potential features for a new personal finance app?"

62. "Suggest names for a new line of vegan leather handbags."

63. "Generate ideas for promoting a new line of men's grooming products."

64. "What are some potential uses for augmented reality in retail?"

65. "Generate ideas for recipes for a food blog focused on gluten-free meals."

66. "What are some strategies for increasing diversity and inclusion in our industry?"

67. "Generate taglines for a new reality TV show about professional chefs."

68. "What are some creative designs for a new line of women's watches?"

69. "Suggest names for a new brand of high-performance running shoes."

70. "Generate ideas for Pinterest boards for a wedding planning business."

71. "What are some potential applications for AI in logistics and supply chain management?"

72. "Generate ideas for workshops for a business coaching company."

73. "What are some ways we could reduce our carbon footprint as a company?"

74. "Generate ideas for Facebook ads for a new line of cruelty-free makeup."

75. "What are some potential features for a new project management tool?"

76. "Suggest names for a new brand of organic baby food."

77. "Generate ideas for blog posts for a website about parenting and early childhood education."

78. "What are some potential uses for blockchain in the music industry?"

79. "Generate ideas for Instagram Reels for a fitness trainer."

80. "What are some strategies for improving employee morale and job satisfaction?"

81. "Generate taglines for a new animated film about wildlife adventures."

82. "What are some creative designs for a new line of minimalist jewelry?"

83. "Suggest names for a new line of eco-friendly office supplies."

84. "Generate ideas for promoting a new podcast about true crime."

85. "What are some potential applications for drones in film production?"

86. "Generate ideas for TikTok videos for a beauty influencer."

87. "What are some ways we could improve the user experience on our mobile app?"

88. "Generate taglines for a new TV series about time travel."

89. "What are some creative ideas for a new line of kids' toys?"

90. "Suggest names for a new brand of organic tea."

91. "Generate ideas for YouTube video topics for a digital marketing expert."

92. "What are some potential uses for AI in journalism?"

93. "Generate ideas for blog posts for a website about home gardening."

94. "What are some strategies for increasing customer retention?"

95. "Generate ideas for LinkedIn posts for a career coaching business."

96. "What are some potential features for a new language learning app?"

97. "Suggest names for a new brand of men's skincare products."

98. "Generate ideas for promoting a new fitness class on social media."

99. "What are some potential uses for virtual reality in tourism?"

100. "Generate ideas for webinars for an online education platform."

101. "What are some ways we could improve the checkout process on our e-commerce website?"

102. "Generate taglines for a new drama film set in the 1920s."

103. "What are some creative designs for a new line of men's wallets?"

104. "Suggest names for a new craft coffee shop."

105. "Generate ideas for podcast episode topics about mental health."

106. "What are some potential applications for blockchain in healthcare?"

107. "Generate ideas for Instagram posts for a personal stylist."

108. "What are some ways we could increase engagement on our company blog?"

109. "Generate ideas for YouTube video topics for a DIY crafts channel."

110. "What are some potential features for a new meal planning app?"

111. "Suggest names for a new line of luxury pet accessories."

112. "Generate ideas for promoting a new restaurant on social media."

113. "What are some potential uses for augmented reality in architecture?"

114. "Generate ideas for recipes for a food blog focused on keto diet."

115. "What are some strategies for reducing employee turnover?"

116. "Generate taglines for a new game show about trivia."

117. "What are some creative ideas for a new line of designer eyewear?"

118. "Suggest names for a new brand of high-end headphones."

119. "Generate ideas for blog posts for a website about personal development."

120. "What are some potential applications for AI in customer service?"

121. "Generate ideas for Instagram IGTV videos for a makeup artist."

122. "What are some ways we could improve our email marketing campaigns?"

123. "Generate ideas for webinars for a digital marketing agency."

124. "What are some potential features for a new graphic design tool?"

125. "Suggest names for a new line of eco-friendly toys."

126. "Generate ideas for promoting a new comedy club on social media."

127. "What are some potential uses for drones in search and rescue operations?"

128. "Generate ideas for Pinterest boards for an interior design business."

129. "What are some strategies for increasing website traffic?"

130. "Generate taglines for a new reality TV show about start-up founders."

131. "What are some creative ideas for a new line of women's athleisure wear?"

132. "Suggest names for a new farm-to-table restaurant."

133. "Generate ideas for blog posts for a website about remote work and digital nomad lifestyle."

134. "What are some potential applications for blockchain in voting and elections?"

135. "Generate ideas for LinkedIn posts for a recruitment agency."

136. "What are some ways we could improve our social media presence?"

137. "Generate ideas for YouTube video topics for a travel blogger."

138. "What are some potential features for a new to-do list app?"

139. "Suggest names for a new brand of non-alcoholic cocktails."

140. "Generate ideas for promoting a new art exhibition on social media."

141. "What are some potential uses for virtual reality in staff training and professional development?"

142. "Generate ideas for webinars for a real estate company."

143. "What are some ways we could improve the user experience on our online booking platform?"

144. "Generate taglines for a new horror film set in space."

145. "What are some creative ideas for a new line of men's watches?"

146. "Suggest names for a new brand of gluten-free snacks."

147. "Generate ideas for podcast episode topics about cryptocurrency."

148. "What are some potential applications for AI in transportation and logistics?"

149. "Generate ideas for Instagram Live sessions for a nutritionist."

150. "What are some ways we could increase user engagement on our forum?"

151. "Generate ideas for YouTube video topics for a gardening channel."

152. "What are some potential features for a new home organization app?"

153. "Suggest names for a new line of vegan protein bars."

154. "Generate ideas for promoting a new online course on social media."

155. "What are some potential uses for augmented reality in education?"

156. "Generate ideas for recipes for a food blog focused on Paleo diet."

157. "What are some strategies for managing remote teams effectively?"

158. "Generate taglines for a new animated series for kids."

159. "What are some creative ideas for a new line of kids' clothing?"

160. "Suggest names for a new coworking space."

161. "Generate ideas for blog posts for a website about mindfulness and meditation."

162. "What are some potential applications for blockchain in real estate?"

163. "Generate ideas for Facebook Live sessions for a local bookstore."

164. "What are some ways we could improve our affiliate marketing program?"

165. "Generate ideas for webinars for an SEO agency."

166. "What are some potential features for a new time tracking app?"

167. "Suggest names for a new brand of plant-based cheeses."

168. "Generate ideas for promoting a new music festival on social media."

169. "What are some potential uses for drones in photography and videography?"

170. "Generate ideas for Pinterest boards for a fashion stylist."

171. "What are some strategies for improving website conversion rates?"

172. "Generate taglines for a new sci-fi film about artificial intelligence."

173. "What are some creative ideas for a new line of eco-friendly furniture?"

174. "Suggest names for a new vegan restaurant."

175. "Generate ideas for blog posts for a website about eco-tourism."

176. "What are some potential applications for AI in retail?"

177. "Generate ideas for Instagram Stories for a celebrity stylist."

178. "What are some ways we could improve our product packaging?"

179. "Generate ideas for YouTube video topics for a tech review channel."

180. "What are some potential features for a new event planning app?"

181. "Suggest names for a new brand of organic wines."

182. "Generate ideas for promoting a new documentary film on social media."

183. "What are some potential uses for virtual reality in interior design?"

184. "Generate ideas for recipes for a food blog focused on diabetic-friendly meals."

185. "What are some strategies for improving online customer reviews?"

186. "Generate taglines for a new adventure game."

187. "What are some creative ideas for a new line of designer handbags?"

188. "Suggest names for a new indie bookstore."

189. "Generate ideas for podcast episode topics about self-care and wellness."

190. "What are some potential applications for blockchain in supply chain management?"

191. "Generate ideas for Instagram posts for a celebrity chef."

192. "What are some ways we could improve our customer referral program?"

193. "Generate ideas for LinkedIn posts for a digital nomad coach."

194. "What are some potential features for a new meditation app?"

195. "Suggest names for a new brand of natural hair care products."

196. "Generate ideas for promoting a new theater play on social media."

197. "What are some potential uses for augmented reality in museum exhibits?"

198. "Generate ideas for TikTok videos for a professional dancer."

199. "What are some strategies for increasing blog subscribers?"

200. "Generate taglines for a new romance novel."

201. "What are some creative ideas for a new line of yoga wear?"

202. "Suggest names for a new brand of sustainable coffee."

203. "Generate ideas for blog posts for a website about green living."

204. "What are some potential applications for AI in banking and finance?"

205. "Generate ideas for Facebook posts for a local farmers' market."

206. "What are some ways we could improve our return and exchange policy?"

207. "Generate ideas for webinars for a graphic design agency."

208. "What are some potential features for a new career coaching app?"

209. "Suggest names for a new brand of essential oils."

210. "Generate ideas for promoting a new art and craft fair on social media."

211. "What are some potential uses for drones in construction and infrastructure?"

212. "Generate ideas for Pinterest boards for a personal organization expert."

213. "What are some strategies for improving mobile app ratings?"

214. "Generate taglines for a new mystery series."

215. "What are some creative ideas for a new line of men's eyewear?"

216. "Suggest names for a new seafood restaurant."

217. "Generate ideas for blog posts for a website about pet care and training."

218. "What are some potential applications for blockchain in cybersecurity?"

219. "Generate ideas for Instagram IGTV videos for a plant-based chef."

220. "What are some ways we could improve our product descriptions

221. "Generate ideas for YouTube video topics for a photography tutorial channel."

222. "What are some potential features for a new fitness tracking app?"

223. "Suggest names for a new brand of herbal teas."

224. "Generate ideas for promoting a new charity event on social media."

225. "What are some potential uses for virtual reality in therapy and mental health treatment?"

226. "Generate ideas for recipes for a food blog focused on vegan cuisine."

227. "What are some strategies for improving customer service?"

228. "Generate taglines for a new animated movie about mythical creatures."

229. "What are some creative ideas for a new line of outdoor adventure gear?"

230. "Suggest names for a new concept cafe."

231. "Generate ideas for blog posts for a website about financial planning and investment."

232. "What are some potential applications for AI in the field of education?"

233. "Generate ideas for Facebook Live sessions for a yoga instructor."

234. "What are some ways we could improve our user onboarding process?"

235. "Generate ideas for webinars for a content marketing agency."

236. "What are some potential features for a new recipe app?"

237. "Suggest names for a new brand of vegan cosmetics."

238. "Generate ideas for promoting a new music album on social media."

239. "What are some potential uses for drones in agriculture and farming?"

240. "Generate ideas for Pinterest boards for a DIY home improvement business."

241. "What are some strategies for improving SEO rankings?"

242. "Generate taglines for a new sci-fi novel about parallel universes."

243. "What are some creative ideas for a new line of sustainable home decor?"

244. "Suggest names for a new vegetarian restaurant."

245. "Generate ideas for blog posts for a website about parenting and family life."

246. "What are some potential applications for blockchain in the music industry?"

247. "Generate ideas for Instagram Stories for a travel influencer."

248. "What are some ways we could improve our customer loyalty program?"

249. "Generate ideas for YouTube video topics for a personal development coach."

250. "What are some potential features for a new music learning app?"

251. "Suggest names for a new brand of organic baby food."

252. "Generate ideas for promoting a new book release on social media."

253. "What are some potential uses for augmented reality in fashion and retail?"

254. "Generate ideas for recipes for a food blog focused on gluten-free meals."

255. "What are some strategies for reducing shopping cart abandonment on our ecommerce site?"

256. "Generate taglines for a new superhero movie."

257. "What are some creative ideas for a new line of ergonomic office furniture?"

258. "Suggest names for a new brand of fair-trade chocolates."

259. "Generate ideas for blog posts for a website about mental health and wellness."

260. "What are some potential applications for AI in the field of healthcare?"

261. "Generate ideas for Instagram IGTV videos for a fitness influencer."

262. "What are some ways we could improve our email marketing campaigns?"

263. "Generate ideas for LinkedIn posts for a career counselor."

264. "What are some potential features for a new e-learning platform?"

265. "Suggest names for a new brand of natural skin care products."

266. "Generate ideas for promoting a new comedy show on social media."

267. "What are some potential uses for drones in emergency services and search and rescue?"

268. "Generate ideas for Pinterest boards for a nutrition coach."

269. "What are some strategies for improving website loading speed?"

270. "Generate taglines for a new historical fiction book."

271. "What are some creative ideas for a new line of sports equipment?"

272. "Suggest names for a new organic juice bar."

273. "Generate ideas for blog posts for a website about entrepreneurship and startups."

274. "What are some potential applications for blockchain in the field of insurance?"

275. "Generate ideas for Facebook posts for a personal finance consultant."

276. "What are some ways we could improve our user interface and user experience?"

277. "Generate ideas for webinars for a cybersecurity consulting firm."

278. "What are some potential features for a new travel booking app?"

279. "Suggest names for a new brand of organic wines."

280. "Generate ideas for promoting a new online course on programming on social media."

281. "What are some potential uses for augmented reality in product demos and virtual tours?"

282. "Generate ideas for recipes for a food blog focused on low-carb diet."

283. "What are some strategies for improving customer satisfaction and retention?"

284. "Generate taglines for a new drama series."

285. "What are some creative ideas for a new line of handmade jewelry?"

286. "Suggest names for a new craft brewery."

287. "Generate ideas for blog posts for a website about sustainable travel."

288. "What are some potential applications for AI in logistics and supply chain management?"

289. "Generate ideas for Instagram posts for a wedding planner."

290. "What are some ways we could improve our online checkout process?"

291. "Generate ideas for YouTube video topics for a language learning channel."

292. "What are some potential features for a new language learning app?"

293. "Suggest names for a new brand of cruelty-free makeup."

294. "Generate ideas for promoting a new art exhibition on social media."

295. "What are some potential uses for drones in wildlife conservation and research?"

296. "Generate ideas for Pinterest boards for a home interior designer."

297. "What are some strategies for managing online reputation?"

298. "Generate taglines for a new fantasy film."

299. "What are some creative ideas for a new line of vegan leather accessories?"

300. "Suggest names for a new artisanal bakery."

301. "Generate ideas for blog posts for a website about minimalism and decluttering."

302. "What are some potential applications for blockchain in the energy sector?"

303. "Generate ideas for Instagram Stories for a lifestyle influencer."

304. "What are some ways we could improve our customer support and service?"

305. "Generate ideas for webinars for a web development agency."

HOW TO USE CHATGPT FOR MARKET RESEARCH (WITH EXAMPLES)

Market research is a critical aspect of business development, marketing, customer attraction and retention, and business growth. It involves identifying your target audience, understanding their pain points and preferences, and understanding the direction of industry trends and competition.

AI language models like ChatGPT can significantly streamline and enhance the market research process, saving you time and money in the process. A lot of times, we don't know how to position our business or our next product or service offering simply because we haven't taken the time to do the market research to give us the data and information to get us to where we need to be.

Identifying market opportunities is a critical aspect of strategic business planning. Here are six ways in which you can use ChatGPT for market research:

1. Gather and Analyze Data

ChatGPT can process and analyze vast amounts of data in a fraction of the time it would take a human to do the same task. It can scan thousands of articles and online content to identify trends, patterns, and key insights relevant to your industry.

For example, you could feed ChatGPT the context and content of articles about the latest technology trends within the e-commerce industry and ask it to summarize the main points for you. This capability allows you to stay updated with the latest developments in your industry and make informed business decisions.

2. Customer Insights and Sentiment Analysis

Understanding who your customers are and what they need in the moment is crucial for any marketing to be successful. With ChatGPT, you can analyze customer reviews from your Google My Business profile, social media comments, and other user-generated content to glean insights into the sentiment and feelings of your customers.

This process could include analyzing customer reviews to identify common complaints, suggestions for improvement, or praise from satisfied customers. It could also include scanning social media comments for feedback on your brand or on your competitors. ChatGPT can assist in sorting through and summarizing this information, helping you gain a clear understanding of your customers' preferences and pain points.

3. Competitor Analysis

Who are your competitors? What are your competitors doing well? What opportunities have your competitors missed? Are their services or product features that customers are looking for currently? ChatGPT can help you answer these questions. Being and staying aware of what your competitors are up to is a key aspect of market research.

ChatGPT can assist in this process by analyzing peer or competitor websites, meta descriptions, product descriptions, and marketing materials. It can summarize their service or product offerings, identify their unique value proposition, and run a comparative analysis between their products or services and your products and services. Additionally, it can help analyze customer reviews and social media sentiment, offering insights into competitor strengths and weaknesses.

4. Survey and Questionnaire Analysis

If you have conducted surveys or questionnaires for your brand or business, ChatGPT can help to analyze the responses. It can sift through the data, identify patterns, and summarize the primary findings making it easier to digest. This can save significant time, money, and resources, allowing you to take quick action on the insights gained from the process.

5. Identify Opportunities in the Market

You can use ChatGPT to identify potential opportunities in the market or industry you serve. For example, you could ask it to analyze data on emerging consumer trends and give you a list of suggestions for new products, services, or improvements to existing offerings. By staying on top of these trends, technology advancements, and even regulatory changes in the market, you can ensure your business remains competitive and responsive to consumer needs.

6. Content Creation and Distribution

Market research isn't just about gathering data and information; it's also about crafting a plan to use the findings and to share insights with your team, stakeholders, and customers. ChatGPT can assist in creating reports, blog posts, presentations, and other content based on your research. It can take raw data and transform it into an engaging, easy-to-understand presentation format or PDF document, making it easier for you and your target audience to digest and act on the information.

7. Diversification and Exploration of New Markets

In some cases, your market research might lead you to expand upon current product or service offerings. If you find that the local or national market is over-saturated with similar offerings, you may want to expand to international markets to broaden your opportunities. The same goes for if your business or brand is growing exponentially locally or nationally, that might give you a hint to explore expanding into other markets that might be in a different region or internationally as well.

While identifying market opportunities is important, equally important is the evaluation of these opportunities. Not every opportunity will be a good fit for your business. It is crucial to assess the potential of each opportunity in light of your business' goals, resources, and capabilities before you decide to pursue it.

Let's say that you are the owner of a local organic skincare company, and you want to analyze another local competitor, who we will call for purposes of this example "Organic Glow." Organic Glow has garnered a bigger market size than you have at the moment, and you want to see what they are doing right to give you some ideas and insights that you can use to improve your marketing strategy. You could use ChatGPT to help in this process.

Here's an example of what your process might look like:

Step 1: Ask ChatGPT to scan the website of Organic Glow and summarize the critical information points. The input might look something like this: *"Here is the text from the About Us page, Product page, and Review Us page of the Organic Glow website. Summarize the mission, product range, marketing strategy, and overall sentiment of their customers."*

Step 2: Ask ChatGPT to analyze their social media profiles and summarize insights on messaging. You may input something like: "Here are the last 25 tweets from Organic Glow. Identify the main themes of their social media messaging."

Step 3: Then, you may want to do something like a SWOT analysis in which ChatGPT will help you to identify potential strengths and weaknesses of the brand. You might ask ChatGPT something like, "Based on the information gathered so far about Organic Glow, give me their top three potential strengths and top three potential weaknesses."

While this is a rather simple illustration of how ChatGPT could help in the competitor analysis and market research process, it is important to keep in mind that ChatGPT should be used in collaboration with other market research tools and methods for the most cohesive and holistic understanding of your competitors.

ChatGPT is an incredibly valuable tool for market research. Understanding prompt engineering and how to tell it what you need so you get the accurate output is a great skill to have here as well. ChatGPT offers a quick, efficient, and cost-effective way to gather and analyze data, understand your customers and competitors, identify market opportunities, and communicate insights. By integrating ChatGPT into your market research process, you can improve your understanding of the current market, make better informed decisions, and ultimately drive your business growth.

HOW TO USE CHATGPT TO BUILD A BUSINESS PLAN (WITH EXAMPLES)

C reating a business plan is a crucial step for many business owners and entrepreneurs in establishing a new venture or expanding an existing one. It is a strategic document that outlines the vision of a company and the steps it plans to take to make that vision a reality. This plan often includes an executive summary, market analysis and projections, financial forecasting, organization and management structure, marketing and sales strategies, and what the plan is to reach a specific goal by a specific timeframe.

On average, to get a business plan written and fashioned for presentation to a business partner, investor, or audience takes several months and can cost tens of thousands of dollars if you bring in a consultant, writer, and designer to help you through the process. Lucky for us, there is ChatGPT that can help us get a very foundational business plan pulled together in a matter of minutes.

Here's how you can use ChatGPT to help develop a comprehensive business plan. Please note, that this is a very basic outline and different types of business plans are needed for businesses that are in different stages of growth.

1. Executive Summary

Start by writing a brief, compelling overview of your business that outlines your business idea, the industry you will operate in, your target market and audience, and what makes your business/product/service/app/idea unique. Use ChatGPT to refine and improve this section by providing it as a draft and asking for suggestions on how to improve it. Or you can simply give ChatGPT a basic sentence about your business and request it to come up with the full executive summary.

For example, you can say, "I am writing a business plan for my new meditation and mental health app that will service the Generation Z population. Give me a 250 word executive summary for my business plan that will explain my company idea and detail what sets us apart in the industry."

2. Company Description

Next, describe your company in detail, including the problems you are providing a solution for, the types of solutions you're offering, and the general business structure. ChatGPT can help guide you on what to include in this section of the business plan.

For example, based on the meditation and mental health app idea above, you can continue the same conversation by saying, "Help me to write a compelling description of my company for my business plan document."

3. Market Analysis

You will want to refer to the previous chapter on market research here. But you can use ChatGPT to help you conduct the basic market analysis, identify your target audience, get insights from competitors, and understand the necessary industry trends that your business partner or investor might want to know.

For example, you can ask ChatGPT to help you identify key insights from other mental health or meditation apps in your space or targeted toward other audiences or to help you understand the mental health trends for the Generation Z population.

4. Organization and Management

This part is a little bit simpler. Here you will want to provide information about the structure of your business, the people on your management or decision-making team, and the roles they play in making your company successful.

ChatGPT can help you to articulate the structure and provide guidance on what to include in this section. For example, with a little bit of information, it can write full bios for you and your team members and give you ideas on what information to include about your team to make your business plan more compelling.

5. Services or Products

Your business plan will need to include information about the services or products you offer. ChatGPT can help you create descriptions and a benefits list of the products or services your business offers, what needs or pain points they meet, and how they stand out in the market.

You will want to ask ChatGPT to help you articulate the benefit of the said product or service and highlight the unique ways that it will help your target audience.

6. Marketing and Sales Strategy

Another key section of a business plan is the marketing and sales strategy – how will you get more people to find out about your product or service and/or make your brand or business more visible. ChatGPT can assist you in outlining your marketing and sales strategy including SEO, content marketing, and social media strategy.

The chapter further in this book on how to create a marketing plan with ChatGPT can help with drafting this information here as well.

7. Financial Projections

While ChatGPT won't be able to generate specific financial projections for you, it can serve as a helpful guide on what type of numbers, percentages, and statistics to include in your business plan and how to present this information.

If you're just starting out, you can ask ChatGPT what information to include in the financial projections section of your business plan. You can also ask it to help you find similar financial projections from competitors to gage your numbers.

8. Request for Funding

If you are seeking funding for your business idea, you will likely need to include a section on how much you're requesting, what the funds will be used for, a proposed repayment plan and what percentage of equity, if any, you will be offering in exchange for the funds. ChatGPT can help you craft a very persuasive and compelling argument for funding.

ChatGPT is always available to help. By interacting with it like this, you can use it to leverage its writing and idea generation capabilities to guide you through each section of your business plan and help you articulate your ideas effectively.

But remember that copying and pasting a business plan from ChatGPT is always a bad idea. You must finesse and edit the copy to ensure it aligns with the goals of your business and that it offers the relevant information that you, a potential business partner or investor would need.

HOW TO USE CHATGPT TO DEVELOP YOUR PRODUCT (WITH EXAMPLES)

Creating a product roadmap is a pivotal element in the product development process. It provides a high-level view of the product's vision and goals and serves as a strategic document that guides the direction of the product team's work. It communicates the why behind what you're building and brings alignment among different stakeholders including developers, marketers, management, and customers.

The product roadmap helps to establish the sequence of product development priorities, allowing teams to understand what features and functionalities need to be built or enhanced, and in what order. It also sets a timeline for the development process, linking strategic objectives with tactical details. By doing so, it assists teams in managing their resources more effectively, balancing the demands of various stakeholders, and keeping the product development on schedule.

In addition, having a good product roadmap documented can act as a tool for tracking the progress of product development over time. It helps monitor whether the team is on track with its goals and allows for timely adjustments as needed. The roadmap also serves as a communication tool that helps the organization understand where the product is headed and what it aims to achieve.

If you are in the product development phase, you likely know that a good product roadmap is not set in stone. It should be a living, evolving document that can adapt to changes in the market, technology, or strategy. It provides a framework that guides the product development process while leaving space for flexibility and adaptation.

In this chapter, we will walk through some ways you can consider using ChatGPT to craft a product roadmap document that works for the initial phases of your product development journey:

1. Product Vision

Your product vision is a long-term, strategic concept of what you want to achieve with your product. ChatGPT can help you refine or give you multiple variations of your product vision to help you get to a version that works for you and your team.

Example: "Our vision is to create an all-in-one fitness app that uses AI technology to provide personalized workout and nutrition plans to help individuals achieve their health and fitness goals."

2. Identify Key Stakeholders

The key stakeholders could include anyone from yourself, your current team members, and your current customers or potential customers to investors and business partners.

Example: "Our key stakeholders include our mobile app users, personal trainers and fitness instructors who provide content for the app, our engineering and development team, and our investors."

3. Set Strategic Goals

This part of the roadmap should define the high-level goals and KPIs you and your team want to achieve throughout this process. ChatGPT can give you some ideas on goals you can set for your product roadmap based on what your product is.

Example: "We plan to increase app downloads by 50% in the first six months. Achieve a user retention rate of 70% and a member referral rate of 30% by the end of the first twelve months. Reach 100% profitability by the end of the second year."

4. Define Product Initiatives

Product initiatives are very high-level efforts, initiatives, or projects that you plan to take up in order to reach or achieve your strategic goals.

Example: "Initiative #1: Develop a user-friendly app interface. Initiative #2: Incorporate 100 workouts and 50 nutrition plans based on body type. Initiative #3: Implement a robust AI algorithm for user personalization."

5. Layout Features and Requirements

This is where you will break down specific initiatives into features and requirements that will be part of the product.

Example: "In Initiative #1, some of the features will include a clean and simple UI, easy navigation, single sign-on feature, and tracking functionality for workout progress."

6. Prioritize Features

Use a method like the RICE scoring system (Reach, Impact, Confidence, Effort) or the MoSCoW scoring system (must-have, should-have, could-have, won't-have) to prioritize certain features in the product.

Example: "Using the RICE scoring system, we've determined that developing a user-friendly interface and incorporating at least 100 workouts to launch are our top two priorities."

7. Create a Timeline

Your timeline should include all the foreseeable stages of your product development process and leave room for review and refinement, plus include the estimated dates for completion of each stage and a tentative date to complete the project.

Example: "Q1: Research and initial development. Q2: Beta testing and refinement. Q3: Launch marketing campaign and official release."

8. Review and Adjust

A product roadmap is a dynamic document that will change as market conditions, customer needs, and business priorities evolve. ChatGPT can help you with continuous analysis of the market, customer trends, and understanding how it fits into the larger picture of your business.

Example: "We plan to review our roadmap biweekly or every quarter to be aligned to significant changes in the market or within our business plan."

Your product roadmap should be a strategic document that communicates the why, what, and how of your overall product development process. It should also be clear enough that anyone in your organization can understand it, take parts of it and run with it, and provide more than enough details to guide your product development team through the development journey. ChatGPT is your friend in this process to help you make changes and get all of the actions and initiatives into words and on paper, cutting the time and energy to do so in half.

Additionally, having a product development roadmap in place can be useful when you bring on business partners or consultants for specific aspects of the project or other external stakeholders so they can see your product's direction and how it will evolve over time.

HOW TO USE CHATGPT TO CRAFT PERSUASIVE MARKETING COPY (WITH EXAMPLES)

E very business, brand, product, or service will require persuasive marketing copy to get and keep your target audience interested in what you are offering.

Crafting persuasive marketing copy involves using compelling language to convince your target audience or potential customers to purchase, invest, read, view, or otherwise be interested in the product or service you are offering.

Not every business starts out with a content writer on their team or even a marketing strategist who can help guide some of these requirements. But with the power of AI models in ChatGPT and other platforms like Writesonic, Content at Scale, Jasper.AI, and Taplio, you can craft compelling, engaging, and persuasive copy that grabs attention and gets users to take action.

Here are five things to consider when you are aiming to write marketing copy that compels, persuades, and converts.

1. Know Your Audience

First things first – you need to really know your audience. Who are they? Where do they hang out? What are their pain points? What are they most concerned about? Why do they need you and what you have to offer? If you know this information, it will help you to input the right type of data into ChatGPT, so it formulates the outputs that resonate with your audience. If you don't know who you are writing for, you will absolutely miss them with the content you produce.

For example, let's say you want to target millennial homeowners with a new interactive checklist. You can ask ChatGPT to generate a list of potential interests, pain points, needs, and concerns that the majority of millennial homeowners have in a given region.

2. Catch Them with the Headlines

Creating clear and catchy headlines is essential for grabbing the reader's attention. In most marketing pieces, people will scan the small copy. If your headlines don't stand out to them, ask them a question, or get them to think quickly about what they might be needing or missing out on, you have lost that potential customer.

Back to our millennial homeowners example above, let's say that you have created a comprehensive guide on mortgage rates that will be useful for this audience. You can ask ChatGPT to give you catchy headline ideas or variations of an idea that you already have. You can also ask it to write you a meta description as well as summary of the article so you can pitch a version of it to more well-known publications.

3. Define the Product or Service

ChatGPT can help you to define what your product or service is or what your business or brand does with language that appeals to a specific audience. Clear definition of your product or service will help people to quickly understand what you do and whether or not you're the perfect fit for them.

Let's say, for example, that you have created a portal of real estate owners who are specialized in helping first-time homeowners. You might ask ChatGPT to give you a description and a list of benefits for the about us page of your website that will help people understand what you do and whether they are in the right place.

4. Focus on the Benefits

Always focus on the benefits of your product or service, not just the features and definitely not just who you are and what you do. Every home buyer needs a real estate agent but why should they pick you over anyone else. Show rather than tell your audience how your product or service will solve their problems or enhance their lives. And let ChatGPT offer ideas on compelling verbiage to use that spotlights the benefits.

Back to our home buying scenario, you can ask ChatGPT to give you the top ten benefits that first-time home buyers can expect by using our portal of real estate agents specialized in helping them buy their first home. You can also ask it to convert the list of features, if you already have these listed, into a list of benefits that matter to the end users.

5. A Strong Call to Action (CTA) Always Helps

Persuasive marketing copy isn't complete without a strong and clear call to action. Your audience needs to know what to do next. Do you want them to sign up for something? Or do you want to them to subscribe to your newsletter or YouTube channel? Whatever action you want the user to take, ChatGPT can give you compelling CTA options that match what you're selling.

Let's say that you're trying to get trial version sign-ups for your first time homebuyer real estate portal, you can ask ChatGPT to give you the best fifteen options for a compelling CTA.

Even when we have so many different tools to help us, writing compelling marketing copy is as much an art as it is a science. While ChatGPT can provide you with a solid foundation and save time, you'll often get the best results by combining its capabilities with your creativity, unique perspective, and knowledge of your business, customers, and goals you want to achieve.

HOW TO USE CHATGPT TO BUILD A CONSISTENT CONTENT CREATION WORKFLOW (WITH EXAMPLES)

C reating a consistent content creation workflow is vital for maintaining a steady output of high-quality content. Every business requires some form of content to share with their audience or stakeholders. The key isn't just to create content but to be consistent with the output and high-quality with the ideas.

As a content creator or marketer, you likely know already how difficult it can be to keep up with a demanding content production schedule. Lucky for all of us, we can take the edge off a little by incorporating ChatGPT and similar AI based tools and platforms into our workflow.

Here are a few things to consider when thinking about streamlining your content creation process and focusing on what your audience needs most and where you need to show up the best.

1. Identify Content Needs

Use ChatGPT to help identify the specific content needs of your business or brand. If you're a beauty consultant, you may focus more on tutorials and how-to videos than on long-form written content. Think about what kinds of content will resonate with your target audience and support your business goals. ChatGPT can help you to decide what types of content your audience might be interested in the most.

2. Content Planning

As we discussed in an earlier chapter, creating a content calendar that helps you to plan and stay organized is a must. ChatGPT can help generate topic ideas, outlines, and length of content and content format recommendations for your brand or business. It can even help you plan a full month, a full three months, or a full year of content at one time.

3. Content Creation

This part can either be very fun or very boring for some of you. If it's boring, get another talented person to assist in actually creating the content if you need it. But you can use ChatGPT to help draft video and podcast scripts, blog outlines and entire first drafts, social media and email newsletter content, and a host of other content types and formats. If you get stuck on introductions, conclusions, calls-to-action and other things like this, ChatGPT can help give variety and options for you.

4. Content Editing

After generating a first draft of your content, you can use ChatGPT to help refine and optimize your content. For example, you can ask it to rewrite certain sections or expand on a section that needs a little bit more attention. Or you can ask it to help you introduce a new section or write a compelling closing argument or thought.

5. Content Distribution

ChatGPT can also help create copy for promoting your content across various platforms. For example, you might not want to post the same caption on Instagram that you do on LinkedIn. You might want to use some content to pitch and other content to post yourself. Let ChatGPT help create the content variations that you need for different channels and platforms.

6. Content Review

Review the performance and engagement of your content and use these insights to inform your future content strategy. Although ChatGPT can't specifically analyze data for you, it can help to brainstorm ways to improve your content and reveal areas where there are gaps based on that analysis.

Let's break down a possible content creation workflow process for a coaching business using ChatGPT. In this example, we'll consider a life coaching business:

1. Identify Content Needs: As a life coach, your content should resonate with individuals seeking guidance on a range of topics or personal growth. You want to use ChatGPT to help you define the content that would resonate with your target audience.

2. Content Planning: Based on the type of content defined in the previous set, ChatGPT can help you create a content calendar based on a cadence that you give it (one blog post every week or 5 social media posts each week).

3. Content Creation: Once you define the content calendar, use ChatGPT to get the initial drafts of your content completed for each channel or platform you plan to publish or post on.

4. Content Editing: Refine, rewrite, optimize, or restructure the initial draft content with ChatGPT's help so it is ready to share with your audience.

5. Content Distribution: Let ChatGPT write the different variations of promotional copy to share across the different platforms. So, you want share videos on Instagram about mindfulness while your posts on LinkedIn may be more direct about your mindfulness program.

6. Content Review: Based on your review of the performance of your content over a period of time, let ChatGPT help you generate suggestions to improve or fill gaps based on those insights.

Remember, a content creation workflow is not one-size-fits-all. What works best for you will depend on your specific business needs and resources. By utilizing AI in your content creation workflow, you'll be able to streamline your process and ensure a consistent output of high-quality, relevant content. The most effective content resonates with your audience's needs and interests while authentically reflecting your unique perspective as a coach. By taking advantage of the capabilities of tools like ChatGPT, you can focus more on the strategic aspects of content creation and less on the nitty-gritty of writing and editing.

PROMPTS AND QUESTIONS YOU CAN USE IN CHATGPT

MARKETING

- Can you provide me with some ideas for blog posts about [topic]?

- Write a 60 second script for an advertisement about [product or service or company].

- Write a product description for my [product or service or company].

- Write a product description for Amazon.com for my [product or service or company].

- Write a description for my book [title here].

- Write a description for my eBook [title here].

- Write a bio for [person's name] and include [current company name] and [characteristics].

- Create [x number of] distinct call-to-action messages and buttons for [product name].

- Give me [x number of suggestions] for inexpensive ways I can promote my [company or product or service] with Google Ads.

- Make a post showcasing the benefits of using our product [product name] for [specific problem or issue].

- Create two Google Ads in an RSA format [using multiple headlines and descriptions] for an A/B test for [product or service or company].

- Give me the steps to develop a marketing strategy for [company or product or service].

- Give me a digital marketing plan for [company or product or service].

- Give me a print marketing plan for [company or product or service].

- How do you create a marketing plan specific to [industry]?

- Write a marketing strategy for [target audience] for [company or service or product].

- Give me [x number] of suggestions to create a digital marketing strategy for [company or service or product].

- Give me [x number] of steps to evaluate a marketing plan for [company].

- Write a 120-character meta description for my Google search ad promoting [product or service or company].

- What marketing channels have the highest ROI?

- Write a 30-character copy for my Google display ad promoting [product or service or company].

- Write a 20-character copy for my [product or service or company] on Google Play.

- Write a 40-character title for [product or service or company] on Google Shopping Ads.

- Write a case study for [company] detailing [topic or product or service] and include [outcome].

- What factors should I consider when quoting for a brand deal with a [type of company] and what ballpark range should I charge? The scope is [list the scope of the project].

- How can I implement a marketing plan on Google Ads?

- How do you create a marketing strategy?

- How to evaluate a marketing plan?

- What is the difference between a digital marketing strategy and traditional marketing strategy?

- What does it mean to implement a marketing plan?

- What are the steps to implement my marketing strategy?

- What is the process to evaluate a marketing plan?

- Write a guide on creating a digital marketing strategy.

- Produce engaging blog articles that align with our content strategy and target our audience's interests and needs.

- Create captivating videos, infographics, and interactive content to enhance user engagement and brand visibility.

- Optimize website content and blog articles for search engine optimization to improve organic search visibility.

- Create a comprehensive digital marketing campaign plan to promote a new product launch across various online channels.

- Analyze website analytics, identify areas for improvement, and implement strategies to optimize website performance and conversion rates.

- Develop a social media content calendar, create engaging posts, and monitor social media analytics and engagement metrics.

SOCIAL MEDIA

- Give me [x number] of social media marketing ideas for [company name] for TikTok.

- Give me [x number] of social media marketing ideas for [company name] for Facebook.

- Give me [x number] of social media marketing ideas for [company name] for Instagram.

- Create an Instagram story idea for my [product/service/company/brand] that will create a sense of anticipation for my [target audience].

- Give me [x number] of social media marketing ideas for [company name] for YouTube.

- Give me [x number] of social media marketing ideas for [company name] for Twitter.

- Give me [x number] of promotional email subject lines for my [company or product or service].

- Give me cold direct message ideas that will engage my [target audience persona] with a unique style on [subject/topic] that will persuade them to take [desired action] on [website/product/service].

- Create a cold direct message that will establish trust and credibility with [target audience] by showcasing the expertise of [brand/business].

- Write a cold direct message that will share the unique value proposition of [product/service/business] and persuade my [target audience] to make a purchase.

- Give me [x number] of social media marketing ideas for [company name] for Pinterest.

- Give me [x number] of social media marketing ideas for [company name] for LinkedIn.

- Give me a social media marketing plan for [company or product or service].

- Give me [x number of suggestions] for inexpensive ways I can promote my [company or product or service] with [social media platform].

- Create a [social media platform] campaign plan for launching [product or service], aimed at [specific target audience].

- Create a social media calendar for [product or service or company] for [x number of days] in a table format, which includes content ideas, post frequency, post timing, and engagement strategies.

- Generate [x number] of creative ways to use Instagram Reels for [product or service or company].

- Generate [x number] of creative ways to use Instagram for [product or service or company].

- How can I maximize the promotional potential of [social media platform] for my [company or product or service] without breaking the bank? Please share [X number of] affordable marketing ideas.

- Give me budget-friendly ways to enhance the visibility of my [company or service or product] through [social media platform]. Could you suggest [x number of] effective techniques?

- Generate [x number] of creative ways to use Facebook for [product or service or company].

- Generate [x number] of creative ways to use TikTok for [product or service or company].

- Generate [x number] of creative ways to use YouTube for [product or service or company].

- Generate [x number] of creative ways to use Twitter for [product or service or company].

- Create an energetic [social media platform name] post to promote [product or service].

- Give me an interesting and engaging question to post on my Facebook Group about [topic].

- Write an engaging tweet on [topic].

- Create an engaging and detailed Twitter thread on [topic] and include relevant hashtags.

- Create [x number] of captions for an Instagram post related to [topic] that will engage the [target audience] and include relevant hashtags and quotes to make the captions stand out.

- Create a promotional LinkedIn post about the benefits of using [product or service].

- Generate [x number] of creative ways to use LinkedIn for [product or service or company].

- Generate [x number] of creative ways to use Threads for [product or service or company].

- Generate [x number] of creative ways to use Pinterest for [product or service or company].

- Write a 100-character description for my [product or service or company] YouTube ad.

- Write sponsored ad copy for Facebook for my [product or service or company] that impacts [specific audience].

- Create a social media caption that targets [specific audience] and explains how our product [product name] can help them.

- Write a list of [x number] of YouTube video ideas for [product or service or company].

- Create [x-number] of calls-to-action taglines for [product/service] that compels [target audience persona] to take [desired action].

BUSINESS

- Analyze the current state of [industry] and include trends, challenges, opportunities, and relevant data and statistics.

- Give me a list of all the top players in [industry or company].

- Provide a guide on managing finances for a small business, including budgeting, cash flow management, and tax considerations.

- What are the most important KPIs for [industry/field].

- Offer a comprehensive guide to small business financing options, including loans, grants, and equity financing.

- Draft a business plan for a startup, including the vision, mission, target market, competition analysis, and financial projections.

- Create a business plan for an online course or digital product, detailing the content, target audience, pricing, marketing strategy, and revenue model.

- Identify recurring keywords or phrases in customer reviews that highlight the strengths or weaknesses of your product or service.

- Give me a short-term and long-term [industry or company] forecast and explain any potential impact of current events or future developments.

- Offer a detailed review of a [software or tool or technology] as it relates to [target audience] for [business].

- Can you provide me with the mathematical formulas for the most important KPIs for [industry/field].

- Generate an example of a transactions dataset that [company] can create.

- Give me the top [x number] of things to do at public networking events.

BRANDING

- Analyze the style, tone and personality of the following text [post text].

- Rewrite this text [post text] using the style, tone and personality of this text [post text].

- Create [type of content] that follows these formatting guidelines [post guidelines].

- Help me develop a brand positioning strategy to differentiate our product in a competitive market.

- Craft compelling brand messaging that aligns with our target audience and resonates with our brand values.

- Analyze brand performance metrics and provide insights on brand awareness and perception.

CONTENT MARKETING/CREATION

- Create a guide that uses content marketing to engage [specific target audience] and rank better on search engines for [website name].

- Create a list of [x number] of LinkedIn articles to write for a [topic].

- Write a [type of blog post] that will engage my [target audience] with a unique perspective on [subject/topic] and persuade them to take [desired action] on [website/product/service].

- Write a [type of blog post] that will tell a compelling story about my [product/service/company] and how it helps [target audience] achieve [desirable goal].

- Create a list of [x number] of articles to write for [website] on [topic].

- Write a comprehensive guide on [topic] for [company] and [target audience].

- Write a whitepaper on [topic] for [company] and [target audience].

- Create a blog title for a listicle article about [product or service] and include these [action words].

- Write [x number of words] about [topic] for [company or product or service] for [specific target audience].

- Write a lesson plan on [topic] for [audience].

- Write a presentation on [topic] for [company or product or service] for [target audience].

- Create a content strategy for [company or product or service].

- Rewrite this content to include [characteristics it should include] and [post the copy to be rewritten].

- Define a content strategy and how businesses can create one.

- Give suggestions on ways to repurpose content.

- What is user-generated content and how can my business use it?

- List some blog title ideas for selling [product] on Amazon.

- List some blog title ideas for selling [product] on Shopify.

- Create a content strategy that aligns with our brand's goals and target audience, including content themes and distribution channels.

- Assess our existing content assets and identify gaps or opportunities for improvement in terms of quality, relevance, and engagement.

- Develop an editorial calendar outlining the content topics, publishing schedule, and key milestones for the upcoming quarter.

- Write a blog post on sustainable fashion trends for the upcoming season.

- Craft engaging product descriptions for a new line of organic skincare products.

- Create a captivating brand story that showcases the mission and values of a social enterprise.

- Create engaging copy for a new product launch that highlights its features, benefits, and unique selling points.

- Craft catchy slogans and taglines for an upcoming marketing campaign to resonate with our target audience.

- Develop a compelling brand story that communicates our values and connects with our target customers emotionally.

SEARCH ENGINE OPTIMIZATION

- List the top [x number] SEO keyword strategies for a new yoga studio.

- Write a 155-character meta description for my blog post about [topic].

- How can I obtain high-quality backlinks to increase the SEO of [website name].

- Provide a list of semantically relevant topics relevant to [topic].

- Give me [x number] of websites I can reach out to for backlinks for [website name].

- Give me a table overview of the local SEO aspects, along with a short description, action list, and their impact, sorted by impact for [website name].

- I want to enhance my website's visibility on Google Images. Can you assist me in optimizing my images and generating image captions that incorporate [keyword]?

- Give me the "people also ask" questions for [topic].

- Create a no follow link for [insert URL].

- Write a list of frequently asked questions for [topic or keyword].

EMAIL MARKETING

- Analyze these metrics to improve email open rates for a [type of company] brand [paste metrics].

- Write an email series on [topic] for [company or product or service] and [target audience]. The email should include [topics the email should include].

- Write follow-up emails to people who attended my [webinar topic] webinar.

- Structure a weekly [newsletter topic] newsletter for my email list [target audience].

- Structure a weekly [newsletter topic] newsletter for my LinkedIn newsletter [target audience].

- Give me [x number] of promotional email subject lines for my [company or product or service].

- Give me cold email ideas that will engage my [target audience persona] with a unique style on [subject/topic] that will persuade them to take [desired action] on [website/product/service].

- Create a cold email that will establish trust and credibility with [target audience] by showcasing the expertise of [brand/business].

- Write a cold email that will share the unique value proposition of [product/service/business] and persuade my [target audience] to make a purchase.

- Give me [x number] of informative email subject lines for my [company or product or service].

- Give me [x number] of salesy email subject lines for my [company or product or service].

- Give me [x number] of subject lines for my [niche] newsletter.

- Give me [x number] of follow-up email subject lines for [company or product or service].

- How can I re-engage inactive subscribers on my email list?

- Please provide tips to improve the deliverability of our weekly [type of newsletter], making sure it lands in the inbox.

- Create a personalized email greeting for a VIP customer.

KEYWORD RESEARCH

- Create a list of [x number] popular questions related to [topic name] that are relevant for [type of business or industry].

- I want to rank for [keyword]. What other keywords should I target?

- My goal is to increase my website's visibility in video search results. Give me some ideas in optimizing my videos and crafting video transcripts that integrate [keyword]?

- I plan to include a roster of keywords in my blog article about [topic]. Could you help me devise a list of pertinent and high-performing keywords to embed in the post?

- Generate a list of [x number] of keyword ideas on [topic] for [company or service or product].

WEB DEVELOPMENT

- Develop an architecture and code for a [website description] website with JavaScript.

- Help me find mistakes in the following code [paste code below].

- Give me ideas on how to structure the website [website description] using WordPress.

- Give me ideas on how to structure the website [website description] using Shopify.

- Give me ideas on how to structure the website [website description] using Squarespace.

- I want to implement a sticky header on my website. Can you provide an example of how to do that using CSS and JavaScript?

- Please continue writing this code for JavaScript [post code below].

- Find the bug with this code: [post code below].

- Generate examples of UI design requirements for a [mobile app].

- Give me examples of website templates for [industry].

- Generate a sample report of a competitor's [website/product/service] by using online customer reviews. For the sake of this analysis, we will focus on product usability.

SALES

- Create a personalized sales email for a [potential customer] who is [what they like or what they do] for my [company] selling [product or service].

- Write a cold email to a prospective customer to introduce them to my [company] and tell them how it can benefit them with [unique selling points].

- Write a sales strategy to improve sales team performance, including the training, incentives, performance metrics, and sales tools.

- Give me [x number] of creative ways to generate leads for my [company].

- Give me [x number] of cross-selling opportunities for my [company or service or product].

- Develop a sales strategy for my [product/service] in the [industry] by asking simple, easy-to-answer questions.

- Give me a summary of my overall sales strategy.

- Write a compelling headline for an ad that grabs attention and introduces [your product/service].

- Create a sales script for my [product/service] in the [industry] by giving me [x-number] of focused questions.

If you need help applying ChatGPT or other AI tools to your business or brand,
email me at connect@dwcreativeconsultingagency.com or DM me your question via social media:
https://www.facebook.com/mdanniwhite7
https://www.instagram.com/mdanniwhite/
https://www.linkedin.com/in/danniwhite/
https://twitter.com/dannimwhite
https://www.tiktok.com/@mdanniw
http://threads.net/mdanniwhite
https://spill.com/p/mdanniwhite

DW
CREATIVE
CONSULTING AGENCY

Looking for marketing strategy and content direction, book a consultation at
www.DWCreativeConsultingAgency.com

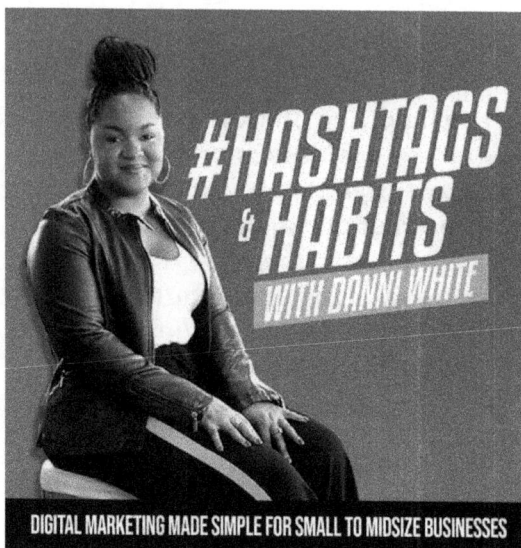

I host a podcast called #Hashtags and Habits where I merge digital marketing, entrepreneurship, and personal development.

Check it out!

Go listen, subscribe, rate, and leave a review to let me know what you think.

Available on Apple Podcasts, Google Podcasts, Stitcher, Spotify, iHeart Radio and YouTube.

www.hashtagsandhabits.com
https://www.instagram.com/hashtagsandhabitspodcast
https://www.facebook.com/hashtagsandhabitspodcast

www.ingramcontent.com/pod-product-compliance
Lightning Source LLC
Chambersburg PA
CBHW071654210326
41597CB00017B/2207